DIFFERENTIAL EQUATIONS IN 24 HOURS

with Solutions and Historical Notes

Scott Imhoff, PhD

Illustrated by Brandon Ross Sugg, BFA
with solutions by Yiran Zhang

The opinions expressed in this manuscript are solely the opinions of the author and do not represent the opinions or thoughts of the publisher. The author has represented and warranted full ownership and/or legal right to publish all the materials in this book.

Differential Equations in 24 Hours
with Solutions and Historical Notes
All Rights Reserved.
Copyright © 2016 Scott Imhoff, PhD
Illustrated by Brandon Ross Sugg, BFA with solutions by Yiran Zhang
V2.0

Cover Photo © 2016 thinkstockphotos.com. All rights reserved - used with permission.

This book may not be reproduced, transmitted, or stored in whole or in part by any means, including graphic, electronic, or mechanical without the express written consent of the publisher except in the case of brief quotations embodied in critical articles and reviews.

Outskirts Press, Inc.
http://www.outskirtspress.com

ISBN: 978-1-4787-6522-6

Library of Congress Control Number: 2015916023

Outskirts Press and the "OP" logo are trademarks belonging to Outskirts Press, Inc.

PRINTED IN THE UNITED STATES OF AMERICA

In memory of Frank Scott Imhoff (1988-2014)

Special thanks to Christopher Riley for help with the manuscript.

Contents

PART 1: **FIRST-ORDER ORDINARY DIFFERENTIAL EQUATIONS** 1

Hour 1. Introduction: What is a Differential Equation? 3

Hour 2. Verifying a Solution 9

Hour 3. Separable Variables 14

Hour 4. Classifying Equations and Solutions 21

Hour 5. Initial Value Problems and Boundary Value Problems 30

Hour 6. Existence and Uniqueness of Solutions 36

Hour 7. Exact Equations 42

Hour 8. Linear Differential Equations of the First Order 49

Hour 9. Homogeneous Coefficients (Method of) 54

Hour 10. Bernoulli's Equation 61

PART 2: **HIGHER-ORDER ORDINARY DIFFERENTIAL EQUATIONS** 65

Hour 11. Euler's Formula 67

Hour 12. Higher-Order Linear ODE's and their IVP's and BVP's 75

Hour 13. Homogeneous Linear ODE's with Constant Coefficients 92

Hour 14. Reduction of Order 100

Hour 15. Undetermined Coefficients 107

Hour 16. Variation of Parameters 112

Hour 17.	Cauchy-Euler Equation	121
Hour 18.	Series Solution	129

PART 3: LAPLACE TRANSFORMS 141

Hour 19.	Introduction to the Laplace Transform	143
Hour 20.	Introduction to the Laplace Transform Exercises	163
Hour 21.	Translation Theorems of Laplace Transforms	165
Hour 22.	Translation Theorems of Laplace Transforms Exercises	183

PART 4: INTRODUCTION TO SYSTEMS OF DIFFERENTIAL EQUATIONS 185

Hour 23.	First-Order Linear Systems with Constant Coefficients	187
Hour 24.	First-Order Linear Systems Exercises	199

APPENDICES

Appendix 1: Answers to Exercises 201

Appendix 2: Timeline 259

Index 261

Preface

The title of this book is intended to be more of a challenge than a promise. No one can promise you that you will learn differential equations in 24 hours. That is up to you. What this book does is it makes it possible for a highly motivated person to learn basic differential equations in the minimum amount of time needed. It has a concise style of presentation and the right number of exercises and examples—not too many, not too few. All of the solutions to all of the exercises are presented in detail in Appendix 1. This allows reinforcement learning and verification of success. The historical material is included because it can be a great source of motivation; however, it can be skipped in the interest of further time savings. The material which can be skipped appears in italics.

Neither is this book intended to be read and the exercises worked in 24 contiguous hours; however, some people have done more in less time. *The Beatles recorded their debut album* Please Please Me *in 13 hours on 11 Feb. 1963. It quickly went to number one. John Lennon, who was sick with a cold, did most of the heavy lifting. Rachel Aaron, a successful author of fantasy and science fiction, writes 10,000 words per day. I did all the work for a course in historical geology at Purdue in a single evening and earned an A.* It takes discipline and desire.

The main prerequisite for the material in this book is integral calculus. This is what is needed to complete the work in Hours 1-22. The material on systems of differential equations, Hours 23 and 24, also requires elementary linear algebra; however, the first 22 Hours comprise what is ordinarily included in a college junior-level course in elementary differential equations. Hours 23 and 24 can be considered bonus material.

The content taught here is equivalent to the material presented in the junior-level course in differential equations that I teach at University of Colorado Denver. It grew out of my earlier book, *Shortcut to Ordinary Differential Equations*. The present book, expanded slightly and equipped with all of the solutions, covers basically the same topics that were taught in a junior-level course in differential equations that I took at Indiana University-Purdue University Indianapolis.

Scott Imhoff, PhD
Parker, Colorado

Part one:
First-order ordinary differential equations

Hour 1.	Introduction: What is a Differential Equation?	3
Hour 2.	Verifying a Solution	9
Hour 3.	Separable Variables	14
Hour 4.	Classifying Equations and Solutions	21
Hour 5.	Initial Value Problems and Boundary Value Problems	30
Hour 6.	Existence and Uniqueness of Solutions	36
Hour 7.	Exact Equations	42
Hour 8.	Linear Differential Equations of the First Order	49
Hour 9.	Homogeneous Coefficients (Method of)	54
Hour 10.	Bernoulli's Equation	61

Hour 1. Introduction: What is a differential equation?

We set about our study of differential equations by starting with an informal definition of what constitutes a differential equation. For present purposes, the following will do:

Definition 1-1: A **differential equation** is an equation involving differentials (e.g., dx, dy, etc.) in which arbitrary constants are suppressed. ∎

An example differential equation is:

$$-x\,dx + dy = 0 \qquad (1\text{-}1)$$

The above is still a differential equation if we manipulate the Leibniz differentials in order to obtain:

$$-x + \frac{dy}{dx} = 0 \qquad (1\text{-}2)$$

Or if we use Lagrange's prime notation to write:

$$-x + y' = 0 \qquad (1\text{-}3)$$

We have to be careful, however, about the latter part of Definition 1-1 regarding arbitrary constants.

An example of what we mean by *arbitrary constant* in the above definition is the familiar C that arises when we evaluate the indefinite integral $\int f(x)dx$ in calculus. Differential equations are written so that these integration constants are not shown. Although derived by integrating Eq. 1-1, the following is not a differential equation since C is not suppressed:

$$-\frac{x^2}{2} + y = C \qquad (1\text{-}4)$$

Already our definition, although simplistic and informal, belies something about the heart of differential equations. Involving differentials and therefore necessarily change, they describe the dynamics of systems. Differential equations are written so as to be the most concise descriptions of general mathematical systems that undergo change. The constants of integration, the C's, which are about particular cases of systems, such as the initial stretch of a spring in a block-spring system, are set aside and the essential meaning of the equation is brought to light. Differential equations bespeak natural and mathematical truths that transcend the particulars of any given application. Differential equations represent broad, overarching generalities about systems that span many types of applications and so express profound insights about generic natural and mathematical systems.

As an example, a stretched spring is just one of many applications described by a very general differential equation known as the *harmonic oscillator*:

$$\frac{d^2y}{dt^2} = -ky \tag{1-5}$$

This equation is simply a description of how the class of system it describes goes about restoring itself when displaced. The equation describes a system that restores itself by accelerating in negative proportion to the displacement: it pulls back without jerk.

As another example, consider the exponential decay equation:

$$y = -b\frac{dy}{dt} \tag{1-6}$$

What this equation describes is the change in the amount of some quantity y. Simply speaking, the rate of change in the amount (of some quantity y) is in negative proportion to how much of y is at hand (at time t). If there is a lot of material, y is decaying rapidly; if there is a small amount, y is decaying more slowly. For such a system, the amount of material y never goes to zero completely.

Both Eq. 1-5 and Eq. 1-6 are compact descriptions of important systems of vastly broad generality. The harmonic oscillator, Eq. 1-5, describes everything from capacitor tuning of an AC circuit to the vibrations of displaced atoms in a crystal lattice[1]. On the other hand, the applicability of the exponential decay model, Eq. 1-6, runs the gamut from describing the rate at which neutrons decay into protons, electrons, and anti-neutrinos[2] to predicting the rate of disappearance of froth in a mug of beer[3].

The distinction between Eq. 1-1 and Eq. 1-4 is that Eq. 1-1 is a differential equation and Eq. 1-4 is its *solution*. To see this, rearrange Eq. 1-4 so that the dependent variable is isolated on one side of the equation:

Hour 1. Introduction: What is a differential equation?

$$y = \frac{x^2}{2} + C \tag{1-7}$$

Thus, while solving a differential equation, ones focus changes from the general to the specific. Once we start dealing with constants of integration, we have passed from the general problem to the specifics and details of a particular case.

Details are important for applications: Being able to solve differential equations empowers you to solve real world problems in technology. Generalities are important too. Often they provide the insights into natural and mathematical systems that inspire true innovation in science. Understanding the general formalism often provides the ability to see similarities in apparently dissimilar systems and this suggests ways to transfer techniques from one problem domain to another. For example, Albert Einstein, through his deep understanding of the harmonic oscillator equation, saw how it could be applied to explain the heat capacity of solids. The first successful application of quantum mechanics to solid state physics was Einstein's Theory of Heat Capacity which was developed by treating each atom in a non-metallic solid as a harmonic oscillator[4]. Thus block-and-spring techniques were applied to a completely different level of problem, the heat capacity, by recognizing that the underlying mathematical structure for both types of systems was the harmonic oscillator.

To gain the full benefit of a course in differential equations, it is necessary to appreciate both the theory, the great generalities the equations represent, and the applications, the solutions to the equations. At the risk of oversimplifying matters, being able to solve differential equations allows one to provide basic engineering solutions. Understanding the deeper insights embodied in the equations themselves allows one to model systems and occasionally to achieve great innovation. We will try to do justice to both viewpoints in this introductory course.

EXERCISES (Calculus Review)

Exercise 1-1:   *Derivative of a rational function.*   Using the rule $\frac{d}{dx}(rq) = q\frac{dr}{dx} + f\frac{dq}{dx}$, show that:

$$\frac{d}{dx}\left(\frac{u}{v}\right) = \frac{v\frac{du}{dx} - u\frac{dv}{dx}}{v^2} \tag{1-8}$$

Differential Equations in 24 Hours

Exercise 1-2: Derivative of tangent. Using the rule for $\dfrac{d}{dx}\left(\dfrac{u}{v}\right)$, determine the following derivative:

$$\frac{d}{dx}(\tan x) \qquad (1\text{-}9)$$

Exercise 1-3: Derivative of secant. Determine the derivative:

$$\frac{d}{dx}(\sec x) \qquad (1\text{-}10)$$

Exercise 1-3: Indefinite integral of secant. Determine the integral:

$$\int \sec x \, dx \qquad (1\text{-}11)$$

Hint: Introduce: $1 = \dfrac{\sec x + \tan x}{\sec x + \tan x}$.

Hour 1. Introduction: What is a differential equation?

Isaac Newton
(1643-1727)
Illustration © 2015, Brandon Ross Sugg

In 1661 Newton entered Trinity College, Cambridge. He received his B.A. degree in January 1664. During the Great Plague (bubonic plague) of 1664-1665, with some recurrence in 1666, Newton had a lot of time on his hands since the university was closed. He used this time to invent calculus, discover the law of universal gravitation, and prove that white light is composed of light of all colors. The notation $\dot{y}$ to represent $\frac{dy}{dt}$ is due to Newton. Newton's second law of motion is a differential equation in the form $\ddot{s} = F(s)$. Newton showed how to calculate the mass of the sun as well as the mass of any planet having a satellite.

In 1696 John Bernoulli and Leibniz proposed a difficult problem known as the brachistochrone problem: Suppose two points are fixed at random in a vertical plane. What is the shape of the curve down which a particle must slide under

Differential Equations in 24 Hours

the influence of gravity so as to pass from the upper point to the lower in the least time? Mathematicians had been working on the problem for six months. Newton came home from a hard day of work, heard of the problem for the first time, and solved it that evening.

Newton was generous with his money, especially to young people.

---

[1]Joseph Callaway, **Quantum Theory of the Solid State**, Academic Press, New York, 1974, p.1.

[2]W. E. Burcham, **Elements of Nuclear Physics**, Longman House, Burnt Mill, U.K., 1981, p. 59.

[3]A. Leike, "Demonstration of the exponential decay law using beer froth," **European Journal of Physics**, Vol. 23, Dec. 2002.

[4]Jennifer Bothamley, **Dictionary of Theories**, Canton, MI: Visible Ink Press, 1993, p.170.

Hour 2. Verifying a solution

A starting point for hands-on work with differential equations is learning to verify solutions of differential equations. A solution of a differential equation is an explicit expression for the dependent variable that makes the equation a true sentence for some specific values of the independent variables. Suppose someone shows you a differential equation and asks if some particular function $y = f(x)$ is a solution of it for some range of the independent variable x. How do you verify that $y = f(x)$ actually works? One approach is to plug $y = f(x)$ into the equation and see if the result is a true sentence for the values of the independent variable(s). Sometimes it's possible to verify that a function $y = f(x)$ is a solution to a differential equation simply by integrating the equation. Thus Eq. 1-4 is a solution to Eq. 1-1.

Example 2-1: Consider the following equation:

$$xy + y' = 0 \qquad (2\text{-}1)$$

A solution, over the reals, is supposed to be:

$$y = Ce^{-\frac{x^2}{2}} \qquad (2\text{-}2)$$

Let's check it:

$$xy + y' = xCe^{-\frac{x^2}{2}} + C\left(-\frac{2x}{2}\right)e^{-\frac{x^2}{2}} = 0 \qquad (2\text{-}3)$$

The sentence is true for all real x; therefore, Eq. 2-2 is a solution to Eq. 2-1. ◊

Example 2-2: Is $y = Cx^2$ a solution to Eq. 2-1 over any interval?

Here we have:

$$y' = C2x \qquad (2\text{-}4)$$

So, plugging into Eq. 2-1, we obtain:

$$xy + y' = xCx^2 + C2x \qquad (2\text{-}5)$$

Differential Equations in 24 Hours

This expression only equals zero, making Eq. 2-1 true, only if $x = 0$. There is no finite interval over which $y = Cx^2$ is a solution. (We might add that expressing the solution as $y = Cx^2$ where x is a *variable* doesn't make sense since x cannot *vary* from 0 and the equation be true.)

EXERCISES

Exercise 2-1: Neutron decay. Given enough time, a neutron n will decay to form a proton p^+, an electron e^-, and an anti-neutrino $\bar{v}$:

$$n \rightarrow p^+ + e^- + \bar{v} \tag{2-6}$$

Suppose y_0 is the starting number of neutrons at time $t = 0$. The time at which only half of the original count of neutrons is present is called the *half-life* and is denoted t_{half}. Verify that:

$$y = y_0 \, e^{-t \frac{\ln 2}{t_{half}}} \tag{2-7}$$

Is a solution to the decay law for radioactive decay:

$$y + \left(\frac{t_{half}}{\ln 2}\right)\dot{y} = 0 \tag{2-8}$$

Exercise 2-2: Harmonic oscillator. Verify that the second-order equation $y'' + y = 0$ has the following two-parameter family of solutions:

$$y = C_1 \cos x + C_2 \sin x \tag{2-9}$$

Exercise 2-3: Special Relativity. A space capsule has length y_0 when it is at rest. As the capsule is brought to a higher and higher speed v, the length y of the capsule--as viewed by an observer at rest-- is a function of the speed v, that is, $y = y(v)$. (Here the length y is considered parallel to the direction of motion of the capsule.) Einstein's Special Theory of Relativity holds that the relationship is:

$$v + k y y' = 0 \quad \text{where} \quad k = \frac{c^2}{y_0^2} \tag{2-10}$$

Hour 2. Verifying a solution

Show that a solution to this equation is:

$$y = y_0\sqrt{1 - \frac{v^2}{c^2}} \qquad (2\text{-}11)$$

Eq. 2-11 is known as a Lorentz Transformation[5]. The constant c is the speed of light.

Exercise 2-4: Determine the value of the constant k for which y(t) a solution to the differential equation:

$$\dot{y} + (\sin 2t)y = 0, \quad y(t) = e^{k\cos 2t}$$

Gottfried Leibniz
(1646-1716)
Illustration © 2015, Brandon Ross Sugg

Differential equations grew out of the calculus and gradually became a separate subject. Isaac Newton solved some differential equations; however, it was Gottfried Leibniz who launched differential equations as a subject in se. Notwithstanding Newton's staggering contributions to knowledge, it was Leibniz who introduced the differential notation dy/dx for the derivative--a notation which was more conducive than Newton's dot notation for developing differential equations.

Gottfried Leibniz was born in 1646, four years after Newton. He received his doctorate in law in 1666. His serious pursuit of mathematics only began when he was 26 and began to be tutored by Christian Huygens. One of Leibniz's early discoveries was the following infinite series:

$$\frac{\pi}{4} = 1 - \frac{1}{3} + \frac{1}{5} - \frac{1}{7} + \frac{1}{9} - \frac{1}{11} + \ldots$$

When Leibniz was 29, he discovered the Fundamental Theorem of Calculus. As Leibniz developed calculus, John and James Bernoulli joined in and by 1704, calculus was an incredibly powerful tool of broad applicability.

Leibniz made major contributions in a number of fields, including philosophy, logic, literature, history, law, religion, and of course mathematics. His day job working for an "elector" required him to travel a lot. Much of his important work was composed as he was bouncing around in a carriage traveling here and there over the cow trails of Europe.

---

[5]Keith R. Symon, **Mechanics**, Reading, Massachusetts: Addison-Wesley Publishing Company, 1971, p.522.

Hour 3. Separable variables

One of Leibniz's first methods for solving differential equations was separable variables; however, although Leibniz had the seed idea for this method, it was John Bernoulli who fully developed and explained the technique.

Suppose a differential equation involves variables y and x and their differentials. To solve a differential equation using separable variables, manipulate the equation so as to place all of the terms involving y together on one side of the equals sign (or on one side of a plus or minus sign) and all of the terms involving x on the other side of the equals sign (or the plus or minus sign). Then integrate.

Example 3-1: Solve the following differential equation:

$$x y + y' = 0 \tag{3-1}$$

Solution: Using Leibniz differential notation we have:

$$x y + \frac{dy}{dx} = 0 \tag{3-2}$$

If we multiply by dx and apply the chain rule we obtain:

$$x y\, dx + dy = 0 \tag{3-3}$$

Dividing through by y (which requires that $y \neq 0$) we obtain:

$$x\, dx + \frac{1}{y} dy = 0 \tag{3-4}$$

Or, segregating the variables to either side of the equals sign we have:

$$\frac{1}{y} dy = -x\, dx \tag{3-5}$$

Hour 3. Separable variables

Integrating, we obtain a solution:

$$\ln(y) = -\frac{x^2}{2} + \ln C \tag{3-6}$$

Exponentiating, we obtain a solution:

$$y = Ce^{-\frac{x^2}{2}} \tag{3-7}$$

◊

Separable variables does not depend on having the variables named x and y nor does it depend on placing all of one variable on one side of the equation (We could have integrated Eq. 3-4 as is.). The separation we require is achieved when the variables, whatever their names are, are sufficiently pulled apart so that solution by integration is immediately available or one or two trivial manipulations away from being integrable.

Definition 3-1: A differential equation exhibits **separable variables** if it can be written in the form:

$$Q(y)\,dy + P(x)\,dx = 0 \tag{3-8}$$

Or:

$$\frac{dy}{dx} = g(x)h(y) \tag{3-9}$$

■

Example 3-2: Solve the differential equation:

$$y' = -y^3 e^x \tag{3-10}$$

Solution: Rewriting using Leibniz differentials, and bringing the right-hand term over to the left we obtain:

Differential Equations in 24 Hours

$$\frac{dy}{dx} + y^3 e^x = 0 \qquad (3\text{-}11)$$

Multiplying through by $y^{-3} dx$ and applying the chain rule, we get:

$$y^{-3} dy + e^x dx = 0 \qquad (3\text{-}12)$$

This is now in the form of Eq. 3-8 and can be integrated to obtain:

$$\frac{y^{-2}}{-2} + e^x = -\frac{1}{2}C \qquad (3\text{-}13)$$

This may be solved for y:

$$y = \frac{1}{\sqrt{C + 2e^x}} \qquad (3\text{-}14)$$

◊

Note that in both of these examples, we chose to write constants of integration in a form convenient for our purposes: In the first example we chose *lnC* and in the second example we chose -½C. We can get away with this since *lnC* can be made to represent any real number. The same is true for -½C.

Example 3-3: Solve the nonlinear first order differential equation:

$$y' - y^2 + 9 = 0 \qquad (3\text{-}15)$$

Solution: We can recast Eq. 3-15 as:

$$\frac{dy}{dx} - (y^2 - 9) = 0 \qquad (3\text{-}16)$$

Applying *dx* and dividing by (y^2 - 9), we have:

$$\frac{dy}{(y - 3)(y + 3)} - dx = 0 \qquad (3\text{-}17)$$

Using partial fractions, this becomes:

$$\frac{\frac{1}{6}dy}{(y-3)} - \frac{\frac{1}{6}dy}{(y+3)} - dx = 0 \qquad (3\text{-}18)$$

Which leads to:

$$\frac{1}{6}\ln(y-3) - \frac{1}{6}\ln(y+3) - x = \ln C \qquad (3\text{-}19)$$

Or:

$$\frac{y-3}{y+3} = Ce^{6x} \qquad (3\text{-}20)$$

This becomes:

$$y = 3\frac{1 + Ce^{6x}}{1 - Ce^{6x}} \qquad (3\text{-}21)$$

◊

There are many types of differential equations for which separable variables will not yield a solution; however, separation of variables is useful often enough for us to include it in a collection of tools for solving differential equations. Separable variables is the first of many tools for solving differential equations that we will gather together in our knowledge base. As we go forward we will add many other tools to our kit. Equipped with any single tool from the collection, we will be only lightly armed to tackle the mainstream of ordinary differential equations; however, once we have collected a dozen or so major tools of the trade, we will be able to hold forth admirably against a variety of problem types.

Differential Equations in 24 Hours

EXERCISES

Exercise 3-1: Treating b as a constant, solve the exponential decay equation by separation of variables:

$$y + b\frac{dy}{dt} = 0 \tag{3-22}$$

Exercise 3-2: Solve the following:

$$\frac{dy}{dx} = y^2 - 4 \tag{3-23}$$

Exercise 3-3: Using separation of variables, solve the relativity equation in which the length of an object y is a function of its velocity v:

$$v + k y y' = 0 \quad \text{where} \quad k = \frac{c^2}{y_0^2} \tag{3-24}$$

(Use the fact that, at $v = 0$, the length is $y = y_0$ to simplify your result.)

Exercise 3-4: Solve:

$$2xy\,dx + x^2\,dy + 6x\,dx = \pi\,dy \tag{3-25}$$

Hour 3. Separable variables

John Bernoulli
(1667-1748)
Illustration © 2015, Brandon Ross Sugg

The Swiss Bernoulli family produced a number of outstanding mathematicians, the first of whom was James Bernoulli (a.k.a. Jacob Bernoulli). John Bernoulli (a.k.a. Johannes Bernoulli) was a younger brother of James Bernoulli. John Bernoulli started out in medicine but soon his mathematical interests overtook him and he learned Leibniz's calculus and began to add to it. John Bernoulli taught calculus to l'Hopital and was well compensated for it by the French mathematician; however, l'Hopital took the notes and published them as a book, taking credit for John's work including l'Hopital's rule which was discovered by Bernoulli.

John could be a difficult person. When his son Daniel (mentioned later in this book) won a prize that John had also competed for, John threw him out of the house.

One of Leibniz's first methods for solving differential equations was separable variables; however, although Leibniz had the seed idea for this method, it was John Bernoulli who fully developed and explained the technique. The harmonic oscillator equation was first recognized and solved analytically by John Bernoulli, in his work modeling a vibrating string.

John Bernoulli had an enormous impact on mathematics and was known as the "Archimedes of his age." An interesting bit of trivia is that one of John Bernoulli's grandsons received his doctorate at age 13.

Hour 4. Classifying equations and solutions

As with any mathematical endeavor, it is useful to classify the problems into different types. This classification helps guide decision making about what types of tools to use to work with the equation. There are several types of differential equations that we will need to distinguish among. There are also a few classes of equations that fall just outside the category of differential equations but which require mention since they are closely related.

This text is concerned primarily with *ordinary differential equations* which can be defined as follows:

Definition 4-1: An **ordinary differential equation (ODE)** is a differential equation in which the derivatives are with respect to one independent variable.

■

The following are examples of ordinary differential equations:

$$\frac{dy}{dx} + 5y = e^x \tag{4-1}$$

$$\frac{d^2y}{dx^2} - \pi\frac{dy}{dx} + 6y = 0 \tag{4-2}$$

Another class of differential equation occurs when multiple partial derivatives of a multivariate function appear. These equations, called *partial differential equations*, are usually not studied in depth until after a course in ordinary differential equations has been completed.

Definition 4-2: A differential equation involving derivatives of a multivariate function with respect to more than one of its independent variables is a **partial differential equation (PDE)**.

■

For a shape $y = y(x, t)$ undergoing melting, a function of position x and time t, the *heat equation* is obeyed. This is a partial differential equation:

$$\frac{\partial^2 y}{\partial x^2} = \frac{\partial y}{\partial t} \tag{4-3}$$

The primed notation for indicating the order of a derivative will often be useful. One prime means one derivative with respect to the independent variable. Two primes mean second derivative. Three primes indicate three derivatives. For four derivatives and higher, we put the number of derivatives in parentheses as a superscript of the function being differentiated:

$$y' \qquad y'' \qquad y''' \qquad y^{(4)} \qquad y^{(5)}$$

first derivative   second derivative   third derivative   fourth derivative   fifth derivative

$$(4\text{-}4)$$

There is another useful notation (of Newton) which involves placing a dot over a variable being differentiated with respect to time. The "fly speck" notation *should only* be used when it's assumed that the differentiation is with respect to time:

$$\dot{y} = \frac{dy}{dt}, \qquad \ddot{y} = \frac{d^2y}{dt^2}, \qquad \dddot{y} = \frac{d^3y}{dt^3} \qquad (4\text{-}5)$$

An important way to classify a differential equation is to identify the order of the equation.

Definition 4-3: The **order** of a differential equation is the highest derivative in the equation (the largest number of times that the dependent variable is differentiated). ∎

The equation below is first order:

$$(y - 2x)dx + 3x\,dy = 0 \qquad (4\text{-}6)$$

The following equation is fourth order:

$$xy^{(4)} + 7y^2 = x^3 \qquad (4\text{-}7)$$

It's easy to be tricked when trying to decide the order of an equation. Try not to be confused between the largest number of times the dependent variable is differentiated (the order) and the largest power that a term is raised to. For

Hour 4. Classifying equations and solutions

example, the equation below is second order (not third order).

$$\left(\frac{d^2y}{dx^2}\right)^3 + 5\frac{dy}{dx} = e^{-x} \qquad (4\text{-}8)$$

There are a couple of ways of writing differential equations that mathematicians find particularly decorous. In the **normal form**, the highest order derivative appears on the left of the equals sign while the lower order items are relegated to the right:

$$\frac{d^n y}{dx^n} = f(x, y, y', \ldots, y^{(n-1)}) \qquad \text{normal form} \qquad (4\text{-}9)$$

In the **general form**, all the terms are arranged as one big function F which is equal to zero:

$$F(x, y, y', \ldots, y^{(n)}) = 0 \qquad \text{general form} \qquad (4\text{-}10)$$

Another important taxonomy of differential equations involves their *linearity*. Linear differential equations are generally much more approachable than nonlinear differential equations.

Definition 4-4: A **linear differential equation** has the property that the derivative of highest order is a linear function of any lower-order derivatives that may be present. ∎

Not surprisingly, with the linearity concept, we also have to speak about the **order of linearity**. Think of a line $z = my + b$ in the y-z plane. When we say that a differential equation is first-order linear, we are only saying that it is *formally* similar to a line such as $z = my + b$. Here dy/dx plays the role of z:

$$\text{First - order linear}: \quad \frac{dy}{dx} = m(x)y + b(x) \qquad (4\text{-}11)$$

For second-order linear differential equations, the second derivative is a linear function of the lower order terms dy/dx and y.

Second - order linear : $\dfrac{d^2 y}{dx^2} = p(x)\dfrac{dy}{dx} + q(x) y + r(x)$

(4-12)

Note that, for this generalized concept of linearity, the "coefficients" m(x), p(x), q(x), and r(x) do not have to be constants. The general n-th order linear differential equation may be written as follows:

n^{th} - order linear :

$$\dfrac{d^n y}{dx^n} = a_{n-1}(x)\dfrac{d^{n-1} y}{dx^{n-1}} + a_{n-2}(x)\dfrac{d^{n-2} y}{dx^{n-2}} +...+ a_0(x) y + a_{-1}(x)$$

(4-13)

Example 4-1: Some examples of linear and nonlinear differential equations. Look at these examples carefully. They may look similar; however, they are drastically different equations.

(4-14)

Linear	Nonlinear
$\dfrac{dy}{dx} + (\sin x) y = 0$	$\dfrac{dy}{dx} + (\sin y) x = 0$
$\dfrac{d^2 y}{dx^2} + x^2 y = 0$	$\dfrac{d^2 y}{dx^2} + x y^2 = 0$

We are now in a position to state a little more carefully what constitutes a solution of a differential equation.

Definition 4-5: The function $y = \phi(x)$ is a **solution** to the equation F(x, y, y', ..., $y^{(n)}$) = 0 on the interval I (on the x-axis) if:

a.) $\phi(x)$ has at least n continuous derivatives.

b.) $\phi(x)$ satisfies the equation (on the interval I).

Hour 4. Classifying equations and solutions

Example 4-2: The equation

$$y + xy' = 0 \qquad (4\text{-}15)$$

Has solution:

$$y = \frac{C}{x}, \qquad x \in I = (-\infty, 0) \cup (0, \infty) \qquad (4\text{-}16)$$

◊

Note how care has been taken to exclude the point $x = 0$ from the domain of the solution.

Many times the flat line $y = 0$ will solve a differential equation. (The example above being a case in point.) The solution $y = 0$ is known as the **trivial solution**. The trivial solution is aptly named because it can often be found by inspection. Also, the trivial solution often represents an unimportant mode of the differential equation (i.e., the block-spring system when the spring is not stretched and everything remains at rest).

In addition to characterizing differential equations, we should say a few words to characterize solutions. An **explicit solution** is one in which the dependent variable is isolated on one side of the equation. In other words $y = \phi(x)$.

Example 4-3:

$$y = \sqrt{1 - x^2}, \quad -1 < x < 1 \qquad (4\text{-}17)$$

Is an explicit solution to:

$$x + yy' = 0 \qquad (4\text{-}18)$$

◊

An **implicit solution** is an equation which solves the differential equation (and does not contain differentials or derivatives) but is not in the form $y = \phi(x)$.

Example 4-4:

$$x^2 + y^2 = 1 \qquad (4\text{-}19)$$

Is an implicit solution to:

$$x + yy' = 0 \qquad (4\text{-}20)$$

◊

Different differential equations may differ in the number of degrees of freedom their solutions have. For an *n*-th order linear differential equation, we will be looking for an **n-parameter family of solutions** rather than a single solution. The first-order differential equation $xy + y' = 0$ has the following one-parameter family of solutions:

$$y = Ce^{-\frac{x^2}{2}} \qquad (4\text{-}21)$$

The second-order equation $y'' + y = 0$ has the following two-parameter family of solutions:

$$y = C_1 \cos x + C_2 \sin x \qquad (4\text{-}22)$$

A **particular solution** is one containing no parameters. The following example is a harmonic oscillator with a forcing function $2 \cos x$.

Example 4-5: Show that function $y = x \sin x$ is a particular solution of $y'' + y = 2\cos x$.

Solution: Differentiating the candidate solution function, we have:

$$y' = \sin x + x\cos x \qquad (4\text{-}23)$$

A second differentiation yields:

$$y'' = \cos x + \cos x - x\sin x = 2\cos x - x\sin x$$

$$(4\text{-}24)$$

Plugging into y″ + y we obtain:

$$y'' + y = 2\cos x - x\sin x + x\sin x = 2\cos x$$

(4-25)

Which shows that the equation $y'' + y = 2\cos x$ is satisfied.

◊

There are some non-differential equations that warrant mentioning. **Difference equations** form the foundation for digital signal processing. An example difference equation is a one-sample delay feedback loop:

$$y[n] = x[n] - y[n-1]$$

(4-26)

The square brackets indicate that the independent variable n assumes integer values. In this course we will learn how to use Laplace transforms to solve higher-order differential equations. The Laplace method that we will be learning is extremely close to a method for solving difference equations called the z-transform method. (In fact the Laplace transform is a special case of the z-transform).

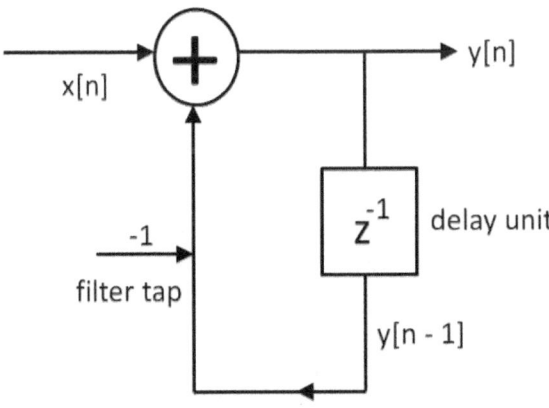

A discrete-time circuit corresponding to Eq. 4-26

Another type of non-differential equation that has come into prominence only recently are dilation equations. They are used to generate functions called *wavelets* which are important for applications ranging from image compression to voice recognition. In dilation equations the independent variables are scaled in the argument more on one side of the equation than on the other. An example dilation equation is the following:

$$\frac{1}{\sqrt{2}}\phi(x) = h_0 \phi(2x) + h_1 \phi(2x) + h_2 \phi(2x) + h_3 \phi(2x)$$

(4-27)

Equations of this type are "solved" using recursion. The solution $\phi(x)$ is "grown" iteratively.

A wavelet function generated using Eq. 4-27.

EXERCISES

Exercise 4-1: Show that the following is a solution to Eq. 4-1 for all x:

$$y = \frac{1}{6}e^x + Ce^{-5x}$$

(4-28)

Exercise 4-2: Show that $y(x, t) = e^{-t}\cos x$ is a solution to Eq. 4-3.

Exercise 4-3: Determine the orders of the following three differential equations:

Hour 4. Classifying equations and solutions

$$\dot{y}^2 + 1 = \dot{y} \tag{4-29}$$

$$(y''')^5 + x^{11}(y^{(4)})^7 = e^{-13x} \tag{4-30}$$

$$x^7(y^3)^{(6)} + (x^7 y^3)^6 + (y''')^{(4)} + e^{-10x} y^{(5)} x^{19} = y^{23} \tag{4-31}$$

Exercise 4-4: Determine which, if any, of the following differential equations are linear differential equations. If an equation is linear, state what order linear it is.

$$y'' + y = 2\cos x \tag{4-32}$$

$$\dot{y} = y^2 + 1 \tag{4-33}$$

$$\frac{d^3 y}{dx^3} - x^7 \frac{dy}{dx} = 0 \tag{4-34}$$

$$y' = x^2 + 1 \tag{4-35}$$

Exercise 4-5: Obtain an explicit solution of the initial value problem and the t-interval of existence.

$$\dot{y} + \frac{1}{y+1} = 0, \; y(1) = 0$$

Hour 5. Initial value problems and boundary value problems

When solving differential equations, we often can do better than to leave the solution in a form involving undetermined constants. When a differential equation is presented together with some additional data about the solution, the added information can often be used to provide a more specific solution. One type of bonus data that is often available is called an initial value. A particular real value of the dependent variable y (or one of its derivatives) at some specific real value of the independent variable t_0 is an **initial value**. The initial value y_0 is a snapshot of "where the system is" at time t_0. Actually, it is correct to call $y_0 = y(t_0)$ an *initial value* even if t_0 is not the starting time of the system. Moreover, the independent variable need not be time t in order for y_0 to be an initial value. The value $y_0 = y(x_0)$ at $x = x_0$ could be an initial value. A differential equation when it is presented together with an initial value of the dependent variable is an initial value problem.

Definition 5-1: An **initial value problem** (abbreviated IVP) is a differential equation and an initial value of the dependent variable (and/or a value of one or more of its derivatives) at some fixed value of the independent variable.

Example 5-1: Bell curve.

If we combine the differential equation $xy + y' = 0$ with the initial value $y(1) = 1$, we have the following initial value problem:

$$IVP: \quad xy + y' = 0 \quad \text{subject to} \quad y(1) = 1 \quad (5\text{-}1)$$

Solution: Use separable variables as in Example 3-1 to obtain:

$$y = Ce^{-\frac{x^2}{2}} \quad (5\text{-}2)$$

Plugging in the initial value $y(1) = 1$, we find that:

$$y(1) = 1 = Ce^{-\frac{(1)^2}{2}} \quad \Rightarrow \quad C = e^{\frac{1}{2}} \quad (5\text{-}3)$$

The solution then is:

$$y = e^{\frac{1}{2}(1 - x^2)} \quad (5\text{-}4)$$

◊

Hour 5. Initial value problems and boundary value problems

Note that the solution has been determined as thoroughly as possible: No parameter C appears here.

For an n-th order differential equation, we will usually require n initial values in order to completely solve the IVP. Thus for the harmonic oscillator, a second order differential equation, two initial values are needed. Often y(0) and y'(0) are given. Note that even if two IVPs have the same differential equation, they are two different problems if their initial conditions are different. The next two examples will illustrate.

Example 5-2: IVP-1; Harmonic Oscillator starting at zero position and upward velocity of 1.

Solve the following initial value problem:

$$IVP-1: \quad \ddot{y} + y = 0 \qquad (5\text{-}5)$$

With initial conditions:

$$y(0) = 0 \quad \text{and} \quad y'(0) = 1$$

A general solution to Eq. 5-5, which will be derived later, is:

$$y = C_1 \cos t + C_2 \sin t \qquad (5\text{-}6)$$

Plugging in the initial values gives us two equations:

$$\begin{array}{lll} y(0) = 0 \Rightarrow & 0 = C_1 \cos(0) + 0 \Rightarrow & C_1 = 0 \\ y'(0) = 1 \Rightarrow & 1 = 0 + C_2 \cos(0) \Rightarrow & C_2 = 1 \end{array} \qquad (5\text{-}7)$$

The solution of IVP-1 is then:

$$y = \sin t \qquad (5\text{-}8)$$

◊

Example 5-3: IVP-2; Harmonic Oscillator starting at position y = 1 with zero velocity. Solve the following initial value problem:

Differential Equations in 24 Hours

$$IVP-2: \quad \ddot{y} + y = 0 \qquad (5\text{-}9)$$

With initial conditions:

$$y(0) = 1 \quad \text{and} \quad y'(0) = 0$$

Again we will work with the (yet-to-be-derived) solution:

$$y = C_1 \cos t + C_2 \sin t \qquad (5\text{-}10)$$

Plugging in the initial values gives us two equations:

$$\begin{aligned} y(0) &= 1 \;\Rightarrow\; 1 = C_1 \cos(0) + 0 \;\Rightarrow\; C_1 = 1 \\ y'(0) &= 0 \;\Rightarrow\; 0 = 0 + C_2 \cos(0) \;\Rightarrow\; C_2 = 0 \end{aligned} \qquad (5\text{-}11)$$

The solution of IVP-2 is then:

$$y = \cos t \qquad (5\text{-}12)$$

◊

The solutions of these two initial value problems are plotted below. (One period is plotted.)

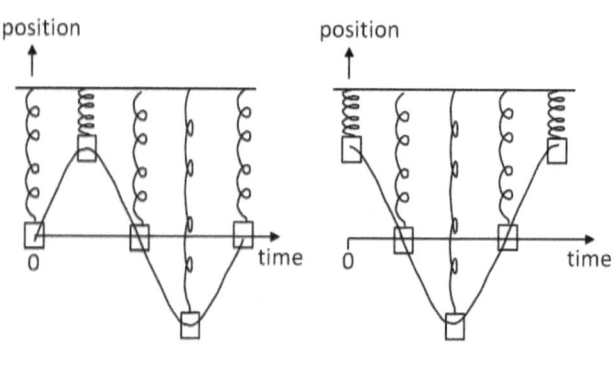

Solution to IVP-1   Solution to IVP-2
(Eq. 5-5)   (Eq. 5-9)

Hour 5. Initial value problems and boundary value problems

The harmonic oscillator equation was first recognized and solved analytically by John Bernoulli, in his work modeling a vibrating string.

In an initial value problem, we have data about the solution (its value and/or the values of one or more of its derivatives) at one fixed value of the independent variable. An equally important type of problem is the **boundary value problem**. Boundary values are values of the dependent variable at two or more *different* values of the independent variable. For example we may have a differential equation and the values of the solution y at two points x_0 and x_1. Then the boundary values are $y_0 = y(x_0)$ and $y_1 = y(x_1)$.

Definition 5-2: A **boundary value problem** (abbreviated BVP) is a differential equation and two or more boundary values. ■

The classical example of a boundary value problem is a spring connected between two walls. At the ends where the spring is nailed down, it cannot move up or down; therefore, in a very literal sense the waves that travel on the spring are bounded. We can think outside the box though and apply the idea of boundary conditions to quite a different situation: Special Relativity.

As was mentioned in Exercise 2-2, in Albert Einstein's Special Relativity, the length of an object, y is a function of the velocity v of the object, that is, $y = y(v)$. As the object goes to higher and higher velocities v the length y becomes shorter and shorter. The celebrated Lorentz contraction is described by the following equation:

$$y = y_0 \sqrt{1 - \frac{v^2}{c^2}} \qquad (5\text{-}14)$$

Where the speed of light is $c = 2.9979 \times 10^8$ m/s and y_0 is the length of the object when it is at rest ($v = 0$). As heady as Eq. 5-14 seems, we can derive it from very simple principles by recognizing a simple boundary value problem. Our starting point will be a relativity equation that simply says that the velocity times its change $v\,dv$ should be in negative proportion to the length times its change $y\,dy$:

$$v\,dv + K\,y\,dy = 0 \qquad (5\text{-}15)$$

There are two boundary conditions to add to this which are conditions on the length y for different values of the independent variable v. The first boundary condition is the simple convention that the length of the object at rest is its original length:

Differential Equations in 24 Hours

$$y(0) = y_0 \qquad (5\text{-}16)$$

The second boundary condition is one of Einstein's postulates. Einstein postulated that for all observers the top speed is $v = c$. Nothing can travel faster than this velocity; therefore, at the speed of light, the object has zero length. This gives our second boundary condition:

$$y(c) = 0 \qquad (5\text{-}17)$$

Thus the Lorentz contraction problem is a boundary value problem. The boundary values are the lengths at the extreme values of speed: 0 and c.

Example 5-4: BVP; Special Relativity.

The boundary value problem for the Lorentz length contraction is:

$$v\,dv + K\,y\,dy = 0 \qquad (5\text{-}18)$$

Subject to:

$$y(0) = y_0 \quad \text{and} \quad y(c) = 0 \qquad (5\text{-}19)$$

Noting that Eq. 5-18 has separable variables, we integrate:

$$\frac{v^2}{2} + K\frac{y^2}{2} = \frac{C_0}{2} \quad \Rightarrow \quad v^2 + Ky^2 = C_0 \qquad (5\text{-}20)$$

Plugging the first boundary value into Eq. 5-19, we have:

$$(0)^2 + K(y_0)^2 = C_0 \qquad (5\text{-}21)$$

The second boundary value, Einstein's postulate, when plugged into Eq. 5-20 gives us C_0:

$$(c)^2 + K(0)^2 = C_0 \quad \Rightarrow \quad C_0 = c^2 \qquad (5\text{-}22)$$

Plugging Eq. 5-22 into Eq. 5-21, we can solve for K as well:

Hour 5. Initial value problems and boundary value problems

$$K = \frac{C_0}{y_0^2} = \frac{c^2}{y_0^2} \qquad (5\text{-}23)$$

Plugging Eq. 5-22 and Eq. 5-23 back into Eq. 5-20 we obtain:

$$v^2 + \frac{c^2}{y_0^2} y^2 = c^2 \qquad (5\text{-}24)$$

Multiplying through by y_0^2 and placing the $y_0^2 v^2$ term on the right, we have:

$$c^2 y^2 = y_0^2 c^2 - y_0^2 v^2 \qquad (5\text{-}25)$$

Dividing by c^2 and taking the positive square root, and factoring out y_0, we obtain:

$$y = y_0 \sqrt{1 - \frac{v^2}{c^2}} \qquad (5\text{-}26)$$

◊

EXERCISES

Exercise 5-1: Solve the following IVP:

$$8t + 2t y + (t^2 - 3)\dot{y} = 0 \qquad (5\text{-}27)$$

With initial value:

$$y(2) = -3 \qquad (5\text{-}28)$$

Exercise 5-2: A space capsule at rest on the launch pad has a length of 10 meters. If the space capsule is brought to a speed that is 80% of the speed of light, what is the length of the space capsule? The speed of light, c, is approximately 3×10^8 meters per second.

35

Hour 6. Existence and uniqueness of solutions

Existence and uniqueness of solutions are subjects that involve a level of rigor that tends to bog down a first course in differential equations; therefore, we will keep this section short. Our main purpose here is to suggest that existence and uniqueness cannot be taken for granted and that for some applications these issues may be vitally important.

Suppose an initial value problem describes the performance of a rocket. A rocket scientist in charge of a manned space launch solves the IVP and concludes that the rocket will take off just fine. Later, as the astronauts prepare for lift off, he becomes nervous and decides to solve the IVP again in order to check his work. This time he determines that the rocket will blow up! Unfortunately he did not make a mistake in either of his calculations: Some differential equations have two or more particular solutions. In other words, for some IVPs, the solution is not **unique**. For critical applications it is important to know about the uniqueness of the solutions of the IVP.

Existence should also not be taken for granted. An example will illustrate the point.

Example 6-1:

Consider the initial value problem:

$$xy' - y \ln y = 0, \qquad y(0) = -1 \qquad (6\text{-}1)$$

Dividing through by $x\,y\,\ln y$, and rearranging slightly, we get:

$$\frac{1}{\ln y}\frac{1}{y}dy = \frac{1}{x}dx \quad \Rightarrow \quad \ln(\ln y) = \ln x + \ln(C)$$

$$(6\text{-}2)$$

Or:

$$y = e^{Cx} \qquad (6\text{-}3)$$

But this result is incompatible with $y(0) = -1$; therefore, there's no solution to the IVP.

36

Hour 6. Existence and uniqueness of solutions

Sometimes we have a solution but it's not unique:

Example 6-2: Consider the initial value problem:

$$y' - xy^{\frac{1}{2}} = 0 \qquad (6\text{-}4)$$

With initial condition:

$$y(0) = 0 \qquad (6\text{-}5)$$

Solution:

The particular solution $y = 0$ solves the equation as well as another particular solution:

$$y = \frac{1}{16}x^4 \qquad (6\text{-}6)$$

So there is not a unique solution.

Picard's Theorem is a tool that we will find useful to help determine whether or not solutions to initial value problems exist and are unique; however, it only applies under the right conditions.

Theorem (Picard):

Consider the following initial value problem:

$$\frac{dy}{dx} = f(x, y) \qquad (6\text{-}7)$$

With initial condition:

$$y(x_0) = y_0 \qquad (6\text{-}8)$$

Let R be the region defined as $\{(x, y) : a \le x \le b,\ c \le y \le d\}$. Let (x_0, y_0) be a point in R. Let $f(x,y)$ be continuous on R and let the partial derivative of f with respect to y, that is, $\frac{\partial f}{\partial y}$, be continuous on R. Then there exists some interval x_0

Differential Equations in 24 Hours

$-h < x < x_0 + h$ (with h positive) in R on which a unique function y(x) solves the initial value problem.

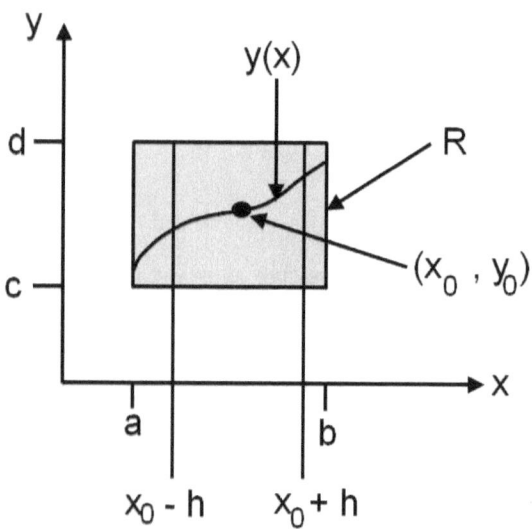

Floor plan of Picard's Theorem

We will not prove the theorem. Mostly we will limit our discussion to the method of it; however, we will say a few words to clear away some of the mystery. The theorem is about how the derivative of y, dy/dx has to be fairly well behaved in order for a solution $y(x)$ of the differential equation to exist. We are given that a solution exists and is unique at x_0 by virtue of having the initial condition $y_0 = y(x_0)$. Picard's Theorem answers the question: What do we need to know about f in order to cultivate this into existence and uniqueness in a larger region.

Understanding the theorem involves a paradigm shift in our thinking. We need to think of the initial condition $y_0 = y(x_0)$ as a <u>point</u> in the x-y plane. The initial condition is the point (x_0, y_0). Any solution $y = y(x)$ to the differential equation must pass through the point (x_0, y_0). In the statement of Picard's Theorem, the requirement on the continuity of f and f_y in the region around (x_0, y_0) force f to have a certain level of good behavior around (x_0, y_0). Since $dy/dx = f(x,y)$, this good behavior of f carries over a certain good behavior for dy/dx which manifests in guarantees of existence and uniqueness of the solution $y(x)$. Using the theorem involves placing the differential equation in the form of Eq. 6-7 and

Hour 6. Existence and uniqueness of solutions

checking to see if the requirements are met: $f(x,y)$ has to be continuous in a neighborhood R of the initial condition (x_0, y_0) and so does the partial derivative $f_y(x,y)$.

Example 6-1: Consider the initial value problem:

$$y' = 3y^{\frac{2}{3}} \quad \text{and} \quad y(0) = 0 \quad (6\text{-}9)$$

Cast this IVP in the form of Eq. 6-6 to see if Picard's theorem applies.

$$\text{Here } (x_0, y_0) = (0, 0) \quad \text{and} \quad f(x,y) = 3y^{\frac{2}{3}} \quad (6\text{-}10)$$

Now the partial derivative with respect to y is:

$$\frac{\partial f}{\partial y} = 2y^{-\frac{1}{3}} \quad (6\text{-}11)$$

Eq. 6-11 "blows up" at (0, 0) since we have a 2/0 cropping up. So Picard's theorem is not met. So we are not guaranteed existence and uniqueness. In this case it turns out that two solutions to this initial value problem are:

$$y = x^3 \quad \text{and} \quad y = 0 \quad (6\text{-}12)$$

Which underscores the failure of uniqueness for this example.

Exercise 6-1: Use Picard's Theorem to see if it guarantees existence and uniqueness for the following IVP:

$$y' - xy^{\frac{1}{2}} = 0 \quad (6\text{-}13)$$

With initial condition:

$$y(0) = 0 \quad (6\text{-}14)$$

Exercise 6-2: Use Picard's Theorem to show that the following IVP has a unique solution:

$$y' + xy = 0 \qquad (6\text{-}15)$$

With initial condition:

$$y(0) = 1 \qquad (6\text{-}16)$$

Hour 6. Existence and uniqueness of solutions

Charles Émile Picard
(1856-1941)
Illustration © 2015, Brandon Ross Sugg

Émile Picard's father died during the siege of Paris in 1870 when Émile was 14. His mother worked hard to provide for Émile and his brother to have a good education. Émile was a brilliant student in almost all subjects but he disliked mathematics. Eventually he read an algebra book and became hooked on math. He received his PhD in mathematics from École Normale Supérieure Paris in 1877. Between 1894 and 1937 Picard trained over 10,000 engineers who were studying at the École Centrale des Arts et Manufactures.

Picard's mathematics resulted in significant contributions to elasticity, electricity, and heat conduction. He used methods of successive approximation to show the existence of solutions of ordinary differential equations. He wrote one of the first textbooks on the theory of relativity.

Picard's father-in-law was the mathematician Charles Hermite, whom we shall meet later. Picard wrote biographies of many leading French mathematicians, including Hermite.

Hour 7. Exact equations

Alexis Claude Cairaut
(1713-1765)
Illustration © 2015, Brandon Ross Sugg

The Englishman Isaac Newton, the German Gottfried Leibniz, and the Swiss Bernoulli brothers founded differential equations in the late 1600's. Later, investigators in other countries began to pick up the torch, the French soon becoming enormously important. Alexis Clairaut was one of the early contributors to differential equations.

Clairaut was a prodigy, studying l'Hopital's calculus book at age 10 and presenting his first paper at age 13 before the Académie française. As a young man, Clairaut was very popular with the ladies and he rarely turned any of them down. He came up with a theorem (Clairaut's theorem) which allowed the calculation of the ellipticity of the Earth.

A team made up of Clairaut, Jérome Lalande and Nicole Reine Lepaute successfully computed the date of the 1759 return of Halley's comet.

Hour 7. Exact equations

We are now ready to add another tool to our box of tools for solving differential equations, one owing to Clairaut and known as *exact equations*.

Remember that, when you have a function of two variables, say, $f = f(x,y)$, the chain rule becomes:

$$df = \frac{\partial f}{\partial x} dx + \frac{\partial f}{\partial y} dy \qquad (7\text{-}1)$$

What does this have to do with ordinary differential equations? It suggests a powerful technique for solving them known as exact equations.

If a differential equation can be written in the form:

$$M(x,y)\,dx + N(x,y)\,dy = 0 \qquad (7\text{-}2)$$

And this can also be expressed as:

$$\frac{\partial f}{\partial x} dx + \frac{\partial f}{\partial y} dy = 0 \qquad (7\text{-}3)$$

Then Eq. 7-2 is an **exact equation**. An exact equation is easy to solve. This is because, by using the chain rule, Eq. 7-3 is the same as saying $df = 0$ or $f = C$. The following theorem is a tool for being able to identify when a differential equation is exact.

Theorem (Clairaut):

Suppose we have a differential equation in the form

$$M(x,y)\,dx + N(x,y)\,dy = 0 \qquad (7\text{-}4)$$

Where M and N have continuous derivatives in x and y. And suppose also that:

$$\frac{\partial M}{\partial y} = \frac{\partial N}{\partial x} \qquad (7\text{-}5)$$

Then Eq. 7-4 is an exact equation.

Note that if we compare Eq. 7-4 and Eq. 7-3, then *exactness* means that:

$$M(x,y) = \frac{\partial f}{\partial x} \quad \text{and} \quad N(x,y) = \frac{\partial f}{\partial y} \qquad (7\text{-}6)$$

Eq. 7-5 is needed because, from elementary calculus, we must have:

$$\frac{\partial^2 f}{\partial y \partial x} = \frac{\partial^2 f}{\partial x \partial y} \qquad (7\text{-}7)$$

If f_x and f_y are continuous. We can determine the solution of Eq. 7-2 by integrating M with respect to x and N with respect to y. We have to be careful to remember that when we integrate with respect to x, we must admit the possibility of a constant function of integration $\phi(y)$. Likewise, when we integrate with respect to y we must realize that a constant function of integration $\psi(x)$ may arise. Then we can reconcile the two integrated equations and hopefully simplify our result.

The algorithm for applying Clairaut's Theorem involves four steps: 1.) Write the equation in $Mdx + Ndy = 0$ form; 2.) Verify Clairaut's condition, i.e., $M_y = N_x$; 3.) Integrate the equations $f_x = M$ and $f_y = N$; 4.) Reconcile the expressions for f and realize $f = C$. The best way to learn is by doing.

Example 7-1: Solve the following:

$$\left(3x^2 y - 2y^3 + 3\right) dx + \left(x^3 - 6xy^2 + 2y\right) dy = 0 \qquad (7\text{-}8)$$

Our first step is to see if the provisos of Theorem 7-1 are met. Looking for the $Mdx + Ndy$ form we recognize that:

$$M(x,y) = 3x^2 y - 2y^3 + 3$$
$$N(x,y) = x^3 - 6xy^2 + 2y \qquad (7\text{-}9)$$

Now the second step is to check for Clairaut's condition, namely, let's see if $M_y = N_x$:

Hour 7. Exact equations

$$\frac{\partial M}{\partial y} = 3x^2 - 6y^2 \quad \text{and} \quad \frac{\partial N}{\partial x} = 3x^2 - 6y^2 \quad (7\text{-}10)$$

So the equation is an exact equation ... or an exact differential.

If you compare:

$$df = \frac{\partial f}{\partial x} dx + \frac{\partial f}{\partial y} dy = 0 \quad (7\text{-}11)$$

With:

$$(3x^2 y - 2y^3 + 3)dx + (x^3 - 6xy^2 + 2y)dy = 0 \quad (7\text{-}12)$$

You can see right away that:

$$\frac{\partial f}{\partial x} = 3x^2 y - 2y^3 + 3 \quad (7\text{-}13)$$

And:

$$\frac{\partial f}{\partial y} = x^3 - 6xy^2 + 2y \quad (7\text{-}14)$$

Watch how these partial derivatives are integrated. This is the third step:

$$f = x^3 y - 2xy^3 + 3x + \phi(y) \quad (7\text{-}15)$$

$$f = x^3 y - 2xy^3 + y^2 + \psi(x)$$

The only way to reconcile these two is to have:

$$\phi(y) = y^2 \quad \text{and} \quad \psi(x) = 3x \quad (7\text{-}16)$$

45

Differential Equations in 24 Hours

So the solution to the differential equation is:

$$f = x^3y - 2xy^3 + 3x + y^2 = C \qquad (7\text{-}17)$$

Or is it? It is important to stop and realize that the original equation we were trying to solve was $M\,dx + N\,dy = 0$ which has nothing to do with f. f is just a helper function; therefore, the correct way to present the final solution is to only report the equation on the right in Eq. 7-17. This completes the fourth step:

$$x^3y - 2xy^3 + 3x + y^2 = C \qquad (7\text{-}18)$$

◊

Example 7-2: Solve the following IVP:

$$2x\sin y - y\sin x + (x^2 \cos y + \cos x)y' = 0 \qquad (7\text{-}19)$$

$$\text{subject to } \frac{\pi}{6} = y\left(\frac{\pi}{2}\right) \qquad (7\text{-}20)$$

Solution: First let's see if the differential equation is exact:

$$M(x,y) = 2x\sin y - y\sin x$$
$$N(x,y) = x^2 \cos y + \cos x \qquad (7\text{-}21)$$

Checking Clairaut's conditions by determining M_y and N_x:

$$\frac{\partial M}{\partial y} = 2x\cos y - \sin x$$

$$\frac{\partial N}{\partial x} = 2x\cos y - \sin x \qquad (7\text{-}22)$$

We see that Clairaut's condition $M_y = N_x$ is satisfied. Thus the equation $M\,dx + N\,dy = 0$ is an *exact differential* and can be expressed as $df = f_x\,dx + f_y\,dy = 0$. This permits us to solve for $f = C$ by integrating. We integrate $f_x = M$ and $f_y = N$:

Hour 7. Exact equations

$$\frac{\partial f}{\partial x} = 2x\sin y - y\sin x$$

(7-23)

$$\frac{\partial f}{\partial y} = x^2 \cos y + \cos x$$

These become:

$$f = x^2 \sin y + y\cos x + \phi(y)$$

(7-24)

$$f = x^2 \sin y + y\cos x + \psi(x)$$

Which can be reconciled by letting $\phi(y) = \psi(x) = 0$. The solution to the differential equation then is, remembering that $f = C$:

$$C = x^2 \sin y + y\cos x$$

(7-25)

But we have not yet solved the IVP. Our last step is to make use of the initial condition $\pi/6 = y(\pi/2)$:

$$C = \left(\frac{\pi}{2}\right)^2 \sin\left(\frac{\pi}{6}\right) + \frac{\pi}{6}\cos\frac{\pi}{2} \Rightarrow C = \frac{\pi^2}{8}$$

(7-26)

So the final solution is:

$$x^2 \sin y + y\cos x = \frac{\pi^2}{8}$$

(7-27)

Differential Equations in 24 Hours

EXERCISES

Exercise 7-1: Use the exact equations method to solve the following:

$$x\,dx + dy = 0 \tag{7-28}$$

Exercise 7-2: Solve the following:

$$3x^2 y\,dx + (x^3 - y^2)dy = 0 \tag{7-29}$$

Exercise 7-3: Solve:

$$(2y + t\cos y)\dot{y} = -\sin y \tag{7-30}$$

Exercise 7-4: Solve the IVP:

$$y\sec^2 x\,dx + \tan x\,dy = 0 \tag{7-31}$$

With initial condition:

$$y\left(\frac{\pi}{4}\right) = 1 \tag{7-32}$$

Exercise 7-5: Solve the following:

$$y\ln y\,dx + x\,dy = 0 \tag{7-33}$$

Hour 8. Linear differential equations of the first order (method of)

We have already learned two tools for solving differential equations, *separable variables* and *exact equations*; we are now ready to learn a third tool. Aptly named, the *method of first order linear equations* applies to the solution of differential equations that are first order linear. In Hour 4 we said that a linear first order ordinary differential equation was one that can be written in the form:

$$\frac{dy}{dx} = m(x)y + b(x) \tag{8-1}$$

And so the 1st order derivative is a linear function of the 0th order derivative. Letting $m(x) = -P(x)$ and $b(x) = Q(x)$, an equivalent form for the linear equation is:

$$\frac{dy}{dx} + P(x)y = Q(x) \tag{8-2}$$

(The method of solution we are about to learn requires the form of Eq. 8-2 rather than the Eq. 8-1 form.)

The method of solution involves making the left-hand side of Eq. 8-2

"look like someone has just applied the product rule." To accomplish this state of affairs, multiply Eq. 8-2 by a certain "integrating factor":

$$e^{\int^{x} P(\psi)d\psi} \tag{8-3}$$

Multiplying Eq. 8-2 by the integrating factor, we have:

$$\frac{dy}{dx} e^{\int^{x} P(\psi)d\psi} + y e^{\int^{x} P(\psi)d\psi} P(x) = Q(x) e^{\int^{x} P(\psi)d\psi} \tag{8-4}$$

But the left hand side is exactly what you obtain when you apply the product

Differential Equations in 24 Hours

rule to $y e^{\int^x P(\psi)d\psi}$:

$$\frac{d}{dx}\left(y e^{\int^x P(\psi)d\psi} \right) = \frac{dy}{dx} e^{\int^x P(\psi)d\psi} + y e^{\int^x P(\psi)d\psi} P(x)$$

(8-5)

Thus Eq. 8-5 can be written:

$$\frac{d}{dx}\left(y e^{\int^x P(\psi)d\psi} \right) = Q(x) e^{\int^x P(\psi)d\psi}$$

(8-6)

Integrating, we see that:

$$y e^{\int^x P(\psi)d\psi} = \int Q(x) e^{\int^x P(\psi)d\psi} dx$$

(8-7)

Then dividing by the integrating factor we can solve for y:

$$y = \frac{1}{e^{\int^x P(\psi)d\psi}} \int Q(x) e^{\int^x P(\psi)d\psi} dx$$

(8-8)

Example 8-1: Solve:

$$x^2 y' + 5xy + 3x^5 = 0$$

(8-9)

And assume that x does not equal zero. Dividing through by x^2 we have:

50

Hour 8. Linear differential equations of the first order (method of)

$$y' + 5\frac{y}{x} = -3x^3 \tag{8-10}$$

This puts it in the proper form so that we can see that:

$$P(x) = \frac{5}{x} \quad \text{and} \quad Q(x) = -3x^3 \tag{8-11}$$

So the integrating factor becomes:

$$e^{\int_{}^{x} P(\psi)d\psi} = e^{\int_{}^{x} \frac{5}{\psi}d\psi} = e^{5\ln x} = x^5 \tag{8-12}$$

Thus the solution is:

$$y = \frac{1}{x^5}\int (-3x^3)x^5\,dx = -\frac{1}{3}x^4 + \frac{C}{x^5} \tag{8-13}$$

◊

Eq. 8-8 probably makes the method look harder than it really is. To use the method it is sufficient to remember to: 1.) Place the equation in the $dy/dx + P(x)y = Q(x)$ form; 2.) Multiply through by the integrating factor $e^{\int P}$; and 3.) Reorganize the left-hand side to be the derivative of a product.

Example 8-2: Solve:

$$x\frac{dy}{dx} - 7y = x^8 e^x \tag{8-14}$$

Divide through by x to get it into the $dy/dx + P(x)y = Q(x)$ form:

$$\frac{dy}{dx} - \frac{7}{x}y = x^7 e^x \tag{8-15}$$

We recognize that $P(x) = -7/x$. The integrating factor then is:

Differential Equations in 24 Hours

$$e^{\int_{}^{x} P(\psi)d\psi} = e^{\int_{}^{x} \frac{-7}{\psi}d\psi} = e^{-7\ln x} = x^{-7} \qquad (8\text{-}16)$$

Multiplying through by x^{-7}, we have:

$$x^{-7}\frac{dy}{dx} - 7x^{-8}y = e^{x} \qquad (8\text{-}17)$$

The left side of this equation is obviously the derivative of $x^{-7}y$.

$$\frac{d}{dx}(x^{-7}y) = e^{x} \qquad (8\text{-}18)$$

Integrating we behold:

$$x^{-7}y = e^{x} + C \qquad (8\text{-}19)$$

Now we can solve for y and we are done!

$$y = x^{7}(e^{x} + C) \qquad (8\text{-}20)$$

◊

EXERCISES

Exercise 8-1: Solve:

$$\frac{dy}{dx} + 2y = e^{2x} \qquad (8\text{-}21)$$

Exercise 8-2: Solve:

$$\cos^{2}x \frac{dy}{dx} + y = e^{-\tan x}\cos^{2}x \qquad (8\text{-}22)$$

Hour 8. Linear differential equations of the first order (method of)

Exercise 8-3: Use the method of first order linear equations to solve:

$$y\,dx + dy - (\sin x)\,dx = 0 \qquad (8\text{-}23)$$

Exercise 8-4: Solve the following:

$$\frac{dy}{dx} + 2xy = e^{-x^2} \qquad (8\text{-}24)$$

Exercise 8-5: An RL circuit initial value problem. In an electric circuit, a constant electromotive force E is provided by a battery. The circuit also contains a resistor R and an inductor L. The current in the circuit as a function of time is $j = j(t)$. The voltage drop across the resistor is jR and the voltage drop across the inductor is $j\,dL/dt$. According to Kirchoff's Law, the sum of these two voltage drops must equal the voltage supplied by the electromotive force E:

$$L\frac{dj}{dt} + Rj = E \qquad (8\text{-}25)$$

The current at time zero is zero, i.e., $j(0) = 0$. Use an integrating factor to solve for the current j as a function of time t: $j = j(t)$.

An RL circuit

Hour 9. Homogeneous coefficients (method of)

Leibniz devised a method for solving equations in the form $M(x,y)dx + N(x,y)dy = 0$ that have a special property for their functions $M(x,y)$, $N(x,y)$ known as homogeneity.

Definition 9-1: A function $f(x, y)$ is **homogeneous of degree** α if, when you replace x with τx and y with τy, you obtain:

$$f(\tau x, \tau y) = \tau^\alpha f(x,y) \qquad (9\text{-}1)$$

■

In other words, the τ "comes through" with power α. Think of homogenized milk and how nicely it flows through a straw. Multipliers which scale the variables "pass through" a function in a comparable way.

Example 9-1:

$$f(x,y) = x^3 + y^3 \qquad (9\text{-}2)$$

Is homogeneous of degree three because:

$$f(tx, ty) = (tx)^3 + (ty)^3 = t^3(x^3 + y^3) = t^3 f(x,y) \qquad (9\text{-}3)$$

◊

Now let's talk about differential equations having *homogeneous coefficients*. You may not be accustomed to thinking about functions as coefficients; however, $M(x,y)$ and $N(x,y)$ are considered to be coefficient functions.

Definition 9-2: A differential equation of the form:

$$M(x,y)dx + N(x,y)dy = 0 \qquad (9\text{-}4)$$

Is said to have **homogeneous coefficients** if M and N are homogeneous functions of the same degree.

■

Hour 9. Homogeneous coefficients (method of)

In order to solve Eq. 9-2, we make a substitution:

$$y = ux \qquad (9\text{-}5)$$

Where u is some function of x. This substitution reduces the above equation to a separable equation. We have the following:

<u>Algorithm for solving 1st order differential equations with homogeneous coefficients:</u>

Step 1.) Test for homogeneity (M and N should be homogeneous of the same degree.)

Step 2.) Substitute $y = ux$.

Step 3.) Solve the separable equation.

Step 4.) Back-substitute to eliminate u (since u is only a helping function).

Let's run this algorithm through its paces.

Example 9-2: Solve:

$$(y^2 - xy)dx + x^2 dy = 0 \qquad (9\text{-}6)$$

First we test for homogeneity:

$$\begin{aligned}M(tx,ty) &= (ty)^2 - (tx)(ty) = t^2(y^2 - xy) = t^2 M(x,y) \\ N(tx,ty) &= (tx)^2 = t^2 x^2 = t^2 N(x,y)\end{aligned}$$

$$(9\text{-}7)$$

So both M and N are homogeneous of degree two. The second step is to substitute $y = ux$. (This will make the differential equation separable.):

$$(u^2 x^2 - xux)dx + x^2(u\,dx + x\,du) = 0 \qquad (9\text{-}8)$$

Which can be written:

Differential Equations in 24 Hours

$$u^2x^2\,dx - x^2u\,dx + x^2u\,dx + x^3\,du = 0 \qquad (9\text{-}9)$$

I can cancel out the x^2:

$$u^2\,dx - u\,dx + u\,dx + x\,du = 0 \qquad (9\text{-}10)$$

The result can now be solved by separable variables:

$$u^2\,dx + x\,du = 0 \qquad (9\text{-}11)$$

$$x^{-1}\,dx = -u^{-2}\,du \qquad (9\text{-}12)$$

$$\ln|x| = \frac{1}{u} - C \qquad (9\text{-}13)$$

The variable u was only introduced to help bridge our way to a solution; therefore, we need to substitute what it represents in order to eliminate it from our result. Thus we substitute u = y/x:

$$\ln|x| = \frac{x}{y} - C \qquad (9\text{-}14)$$

We can simplify this and present our final answer:

$$y = \frac{x}{\ln|x| + C} \qquad (9\text{-}15)$$

◊

It's very straightforward to prove that the algorithm should work.

Theorem: First order homogeneous coefficient equations are reducible to separable

If $M(x,y)dx + N(x,y)dy = 0$ then, if $M(x,y)$ and $N(x,y)$ are homogeneous of the same degree, the equation can be made separable by the substitution $y = ux$.

Proof: Make the substitution $y = ux$:

Hour 9. Homogeneous coefficients (method of)

$$M(x,ux)\,dx + N(x,ux)(x\,du + u\,dx) = 0 \quad (9\text{-}16)$$

Then the homogeneity means that x^α will factor out:

$$x^\alpha M(1,u)\,dx + x^\alpha N(1,u)(x\,du + u\,dx) = 0 \quad (9\text{-}17)$$

Dividing by x^α, the equation is then:

$$M(1,u)\,dx + N(1,u)x\,du + N(1,u)u\,dx = 0 \quad (9\text{-}18)$$

Or:

$$(M(1,u) + uN(1,u))\,dx + N(1,u)x\,du = 0 \quad (9\text{-}19)$$

Dividing through by these coefficient functions, we achieve separation of variables:

$$\frac{1}{x}\,dx + \frac{N(1,u)}{M(1,u) + uN(1,u)}\,du = 0 \quad (9\text{-}20)$$

Q.E.D.

Sometimes it takes effort to overcome changes in notation; however, the methods we are learning do not depend on any particular variable names. In the following the dependent variable is x and the independent variable is t.

Example 9-3: Solve the IVP:

$$\dot{x} + 1 + \frac{9}{t}x = 0$$

With initial condition: $x(10) = 1023 \quad (9\text{-}21)$

Here we are looking for a $M(t, x)dt + N(t, x)dx = 0$ form. If we multiply through by t and dt, we're one step closer:

$$t\,dx + (t + 9x)\,dt = 0 \quad (9\text{-}22)$$

57

Differential Equations in 24 Hours

We need only switch the order of the two terms and we're in $M(t, x)dt + N(t, x)dx = 0$ form.

$$(t + 9x)dt + t\,dx = 0 \qquad (9\text{-}23)$$

Now we see that the coefficient functions are homogeneous of degree one:

$$M(\tau x, \tau t) = (\tau t + 9\tau x) = \tau(t + 9x) = \tau M(x, t) \qquad (9\text{-}24)$$

$$N(\tau x, \tau t) = \tau t = \tau N(x, t)$$

We have to be careful about the present change in notation. The substitution we need to make is $x = u\,t$. This is because Eq. 9-5 really tells us to replace the dependent variable with u times the independent variable. In the present problem, the dependent variable happens to be x and the dependent variable t. Substituting, we obtain $x = u\,t$:

$$(t + 9ut)dt + t\left(u\,dt + t\,du\right) = 0 \qquad (9\text{-}25)$$

We can cancel a t out of every term:

$$(1 + 9u)dt + u\,dt + t\,du = 0 \qquad (9\text{-}26)$$

Next we can use separable variables:

$$(1 + 10u)dt + t\,du = 0 \qquad (9\text{-}27)$$

This becomes:

$$\frac{1}{t}dt + \frac{1}{1+10u}du = 0 \qquad (9\text{-}28)$$

Integrating:

$$\ln|t| + \frac{1}{10}\ln|1+10u| = \ln C \qquad (9\text{-}29)$$

Hour 9. Homogeneous coefficients (method of)

Or:

$$t(1 + 10u)^{1/10} = C \tag{9-30}$$

We now take out the *helper function* $u = x/t$:

$$t(1 + 10\frac{x}{t})^{1/10} = C \tag{9-31}$$

The initial condition says that when $t = 10$, we have $x = 1023$. This lets us solve for C:

$$10(1 + 10\frac{1023}{10})^{1/10} = C \implies C = 10(1024)^{1/10} = 20$$

$$\tag{9-32}$$

The solution then is:

$$x = 2\left(\frac{20}{t}\right)^9 - \frac{t}{10} \tag{9-33}$$

◊

EXERCISES

Exercise 9-1: Solve the following:

$$(x^2 + y^2)\,dx + 2xy\,dy = 0 \tag{9-34}$$

Exercise 9-2: Solve the following:

$$\left(x\cos\frac{y}{x} - y\right)dx + x\,dy = 0 \tag{9-35}$$

Exercise 9-3: Solve:

$$(x + 3y)dx + x\,dy = 0 \qquad (9\text{-}36)$$

Exercise 9-4: Solve:

$$x - \left(t + x\tan\frac{t}{x}\right)\dot{x} = 0 \qquad (9\text{-}37)$$

Hour 10. Bernoulli's equation

Bernoulli's equation has the basic form:

$$\frac{dy}{dx} + \psi(x)y = \phi(x)y^n \tag{10-1}$$

This can be solved by making the substitution:

$$u = y^{1-n} \tag{10-2}$$

This substitution then reduces the differential equation to a separable form.

Sometimes it's easier to solve the problem by multiplying through by $(1-n)y^{-n}$ first before making the substitution. This makes it easy to recognize u and its derivative and to substitute in order to eliminate y. Multiplying Eq. 10-1 by $(1-n)y^{-n}$, we obtain:

$$(1-n)y^{-n}y' + (1-n)\psi(x)y^{1-n} = (1-n)\phi(x) \tag{10-3}$$

Now making the substitutions $(1-n)y^{-n}y' = u'$ and $y^{1-n} = u$, we have:

$$u' + (1-n)\psi(x)u = (1-n)\phi(x) \tag{10-4}$$

The equation is now first order linear and can be solved using an integrating factor:

$$e^{\int^{x}(1-n)\psi(\xi)d\xi} \tag{10-5}$$

Example 10-1: Solve the following:

$$y' + (\tan x)y = \frac{1}{\cos^3 x}y^5 \tag{10-6}$$

In this case $n = 5$. Multiplying through by $(1-5)y^{-5}$, we have:

$$-4y^{-5}y' - 4(\tan x)y^{-4} = \frac{-4}{\cos^3 x} \quad (10\text{-}7)$$

We make the substitution $u = y^{1-5} = y^{-4}$:

$$u' - 4(\tan x)u = \frac{-4}{\cos^3 x} \quad (10\text{-}8)$$

This is now first order linear. We may determine the integrating factor as follows:

$$e^{-4\int^x \tan \xi\, d\xi} = e^{4\int^x \frac{1}{\cos \xi}(-\sin \xi)\, d\xi} = e^{4\ln(\cos x)} = \cos^4 x \quad (10\text{-}9)$$

Applying this integrating factor, we observe:

$$\left(\cos^4 x\right)u' + 4\cos^3 x\left(-\sin x\right)u = -4\cos x \quad (10\text{-}10)$$

Gathering together the left side into the derivative of a product, we have:

$$\frac{d}{dx}\left(u \cos^4 x\right) = -4\cos x \quad (10\text{-}11)$$

This is straightforward to integrate:

$$u\cos^4 x = -4\sin x + C \quad (10\text{-}12)$$

Substituting to eliminate u and solving for y, we have:

$$y = \frac{\cos x}{\left[C - 4\sin x\right]^{\frac{1}{4}}} \quad (10\text{-}13)$$

◊

EXERCISES

Exercise 10-1: Solve:

$$xy' + 2y = 4x^4y^4 \qquad (10\text{-}14)$$

Exercise 10-2: Solve:

$$xy^3y' + y^4 = \cos(x^4) \qquad (10\text{-}15)$$

Exercise 10-3: Solve:

$$\frac{dy}{dx} + \frac{1}{x}y = y^3\cos\left(\frac{1}{x}\right) \qquad (10\text{-}16)$$

Part two:
Higher-order ordinary differential equations

Hour 11.	Euler's Formula	67
Hour 12.	Higher-Order Linear ODE's and their IVP's and BVP's	75
Hour 13.	Homogeneous Linear ODE's with Constant Coefficients	92
Hour 14.	Reduction of Order	100
Hour 15.	Undetermined Coefficients	107
Hour 16.	Variation of Parameters	112
Hour 17.	Cauchy-Euler Equation	121
Hour 18.	Series Solution	129

Hour 11. Euler's formula

Leonhard Euler
(1707-1783)
Illustration © 2015, Brandon Ross Sugg

The Swiss mathematician Leonhard Euler was possibly the greatest mathematician who ever lived. He contributed to many areas of mathematics including calculus, number theory, and graph theory. He did his PhD under John Bernoulli. He generated about 800 pages of new mathematics per year, including some of math's most magnificent discoveries. During the last 17 years of his life, he was totally blind; however, his output actually increased. He had reached a level where he no longer needed to see; he could keep track of vast blackboards of mathematics in his mind.

He was an exceptionally devout Christian who studied theology, Greek and Hebrew in order to become a pastor before his mathematical interests overtook him. It was his Christian faith that helped him maintain his high spirits as he became totally blind. He would write his equations using chalk on a large slate and dictate to his sons what words to put with the equations.

Differential Equations in 24 Hours

The use of the symbol e for the inverse-log of one, the use of π to express the ratio of a circle's circumference to its diameter, and the use of i to represent the square root of negative one, are all notations introduced by Euler. Euler's formula expresses the complex exponential sinusoid $e^{i\theta}$ as the sum of a real summand $\cos\theta$ and an imaginary summand $i\sin\theta$.

As an interesting point of trivia, a granddaughter of Euler married John Bernoulli's grandson Jacob.

Euler's formula is as follows:

$$e^{i\theta} = \cos\theta + i\sin\theta$$
$$\text{where } i = \sqrt{-1} \text{ and } e = 2.718281828...\quad (11\text{-}1)$$

Numbers in the form $a + ib$, where a and b are real, are called **complex numbers**. Real numbers have the property that $b = 0$. Pure **imaginary numbers** have the property that $a = 0$. **Gaussian integers** have the property that a and b are integers. If a, b, c, and d are real numbers, we can define complex numbers z and w as follows:

$$z = a + ib$$
$$w = c + id \quad (11\text{-}2)$$

To add two complex numbers, we proceed as we are used to, only being careful to keep the real terms together and the imaginary terms together:

$$z + w = a + c + i(b + d) \quad (11\text{-}3)$$

To multiply complex numbers we proceed using the approach of multiplying two sums, being careful to substitute $i^2 = -1$ before presenting the result:

$$zw = (a + ib)(c + id)$$
$$zw = ac + i^2 bd + iad + ibc$$
$$zw = ac - bd + i(ad + bc) \quad (11\text{-}4)$$

Hour 11. Euler's formula

Multiplying a complex number by a scalar k has the scaling effect we expect:

$$kz = k(a + ib) = ka + ikb \qquad (11\text{-}5)$$

Multiplying a complex number by an imaginary number ik works as follows:

$$ikz = ik(a + ib) = ika + i^2 kb = -kb + ika \qquad (11\text{-}6)$$

The **conjugate** of a complex number z, denoted z^*, is formed by replacing i with $-i$ everywhere it occurs.

$$(a + ib)^* = a - ib \qquad (11\text{-}7)$$

If a function $f(x, y)$ involves i, then $f(x, y)$ is a **complex function**. The **conjugate of a complex function** $f^*(x,y)$ is formed by replacing every i with $-i$ everywhere it occurs. Let's conjugate a complex exponential to illustrate:

$$h(t) = e^{i\frac{2\pi}{1024}t} \qquad (11\text{-}8)$$

$$h^*(t) = e^{-i\frac{2\pi}{1024}t} \qquad (11\text{-}9)$$

The **norm of a complex number** z, denoted $\|z\|$ is equal to the square root of z times its complex conjugate:

$$\|z\|^2 = z^* z \qquad (11\text{-}10)$$

Euler's formula can be used to prove all sorts of interesting facts about complex numbers. For instance it is possible to raise the square root of negative one to the square root of negative one power using Euler's formula. The result is a real number:

Differential Equations in 24 Hours

Example 11-1: Reduce i^i to a real number.

Plug $\theta = \pi/2$ into Euler's formula:

$$e^{i\frac{\pi}{2}} = \cos\frac{\pi}{2} + i\sin\frac{\pi}{2} = i \qquad (11\text{-}11)$$

This gives an expression for i:

$$i = e^{i\frac{\pi}{2}} \qquad (11\text{-}12)$$

Raise each side of this equation to the i power:

$$i^i = \left(e^{i\frac{\pi}{2}}\right)^i = e^{i^2 \frac{\pi}{2}} = e^{-\frac{\pi}{2}} \approx .21 \qquad (11\text{-}13)$$

So i raised to the i is about one fifth. The solution is not unique.

It's easy to prove Euler's theorem if one makes note of the fact that the powers of i follow the pattern:

$$
\begin{aligned}
i^0 &= 1 \\
i^1 &= i \\
i^2 &= -1 \\
i^3 &= -i \\
i^4 &= 1 \\
i^5 &= i \\
i^6 &= -1 \\
i^7 &= -i \\
&\vdots
\end{aligned}
\qquad (11\text{-}14)
$$

Hour 11. Euler's formula

The Taylor expansions for $\cos(\theta)$ and $\sin(\theta)$ are:

$$\cos(\theta) = 1 - \frac{(\theta)^2}{2!} + \frac{(\theta)^4}{4!} - \ldots$$

$$\sin(\theta) = \theta - \frac{(\theta)^3}{3!} + \frac{(\theta)^5}{5!} - \ldots$$

(11-15)

If we multiply $\sin(\theta)$ by i and add it to $\cos(\theta)$, we have, interleaving the terms:

$$\cos(\theta) + i\sin(\theta) =$$
$$1 + i\theta - \frac{(\theta)^2}{2!} - \frac{i(\theta)^3}{3!} + \frac{(\theta)^4}{4!} + \frac{i(\theta)^5}{5!} \ldots$$

(11-16)

Now using the identities in Eq. 11-14 we have:

$$\cos(\theta) + i\sin(\theta) =$$
$$1 + i\theta + \frac{(i\theta)^2}{2!} + \frac{(i\theta)^3}{3!} + \frac{(i\theta)^4}{4!} + \frac{(i\theta)^5}{5!} \ldots$$

(11-17)

But this is the Taylor expansion for $e^{i\theta}$ which proves Euler's formula:

$$e^{i\theta} = \cos\theta + i\sin\theta$$

(11-18)

◊

Example 11-2: Determine the cube roots of negative one.

Plugging $\theta = \pi$ into Euler's formula we have:

$$e^{i\pi} = \cos\pi + i\sin\pi = -1$$

(11-19)

Differential Equations in 24 Hours

But since cos( ) and sin( ) are both periodic with period 2π, the same is true for $e^{i(\ )}$. This means we can always add $2\pi n$ to the argument without impunity:

$$-1 = e^{i(\pi + 2\pi n)} \tag{11-20}$$

To find the cube roots, we raise each side to the one-third:

$$(-1)^{1/3} = \left(e^{i(\pi + 2\pi n)}\right)^{1/3} = e^{i(\frac{\pi}{3} + \frac{2}{3}\pi n)} \tag{11-21}$$

As n assumes integer values 0, 1, 2, ..., we see that there are three cube roots and then the pattern repeats. The three distinct cube roots are:

$$(-1)^{1/3} = \begin{cases} e^{i(\frac{\pi}{3})} \\ e^{i(\pi)} \\ e^{i(\frac{5}{3}\pi)} \end{cases} \tag{11-22}$$

These can be reduced using Euler's formula:

$$(-1)^{1/3} = \begin{cases} \dfrac{1}{2} + i\dfrac{\sqrt{3}}{2} \\ -1 \\ \dfrac{1}{2} - i\dfrac{\sqrt{3}}{2} \end{cases} \tag{11-23}$$

◊

The **inner product** between two functions g(t) and h(t) having the same domain $-L \leq t \leq L$, is formed by multiplying the first function times the conjugate of the second and integrating the result:

$$g(t) \cdot h(t) = \int_{-L}^{L} g(t) h^*(t) \, dt \tag{11-24}$$

Functions that have a zero inner product are **orthogonal functions**.

Hour 11. Euler's formula

Example 11-3: The two functions sin(t) and cos(t) are orthogonal functions, since their inner product is zero:

$$\int_{-L}^{L} \cos(t)\sin(t)\,dt = -\frac{\cos^2(t)}{2}\bigg|_{-L}^{L} = 0 \tag{11-25}$$

◊

EXERCISES

Exercise 11-1: Calculate the norm of 3 + 4i:

Exercise 11-2: Calculate the following:

$$\cos\left(\frac{\ln i}{i}\right) \tag{11-26}$$

Exercise 11-3: Determine the fourth roots of one.

Exercise 11-4: What is the product of the right hand side of Eq. 11-8 and the right hand side of Eq. 11-9?

Exercise 11-5: The *norm of a function* is the square root of the inner product of the function with itself:

$$norm^2 = \|f(t)\|^2 = \int_{-L}^{L} f(t)f^*(t)\,dt \tag{11-27}$$

What is the norm of h(t), the right hand side of Eq. 11-8? Use L = π.

Exercise 11-6: Very important in nearly all types of engineering, the Fourier Transform of a function f(t) is formed by taking the inner product between the function f(t) and the *Fourier kernel* h(t):

$$h(t) = e^{i\omega t} \tag{11-28}$$

73

Differential Equations in 24 Hours

Letting $L = \pi$, what is the Fourier transform of $f(t) = e^{i\omega t}$?

Exercise 11-7: 0.21 is only one solution of i^i. Let $i = \exp(i(\pi/2 + 2\pi n))$ and find at least two other solutions.

Exercise 11-8: Is 5 really a prime number? Factor 5 into the product of two Gaussian integers.

Hour 12. Higher-order linear differential equations and their IVP's and BVP's

When we solve problems involving higher-order differential equations, we will want to work from the general to the specific: At first we will determine a solution to the differential equation that is as general as possible. Thus we will use just as many free parameters as are needed in order to express every possible manifestation of a solution function to the differential equation. For example, for the differential equation $\ddot{u} + u = 0$, the function $u(t) = C_1\cos(t) + C_2\sin(t)$ is the **general solution**: There is no solution that cannot be expressed by some choice of C_1 and C_2. After the general solution has been determined, we finish solving the problem by using initial conditions or boundary conditions in order to find as many specific values for the C's as possible. Ideally, if things work in our favor, we determine a unique solution. For important types of differential equations, we can determine general solutions by recognizing the set of solutions as a *solution space*--a vector space whose vectors are solution functions to the differential equation. Then, if the differential equation has the right properties, and if we have enough initial or boundary conditions, we can determine the unique point (C_1, C_2) in the solution space that represents the unique solution. To warm up to this idea, we will start by discussing the solution of $\ddot{u} + u = 0$ as a sort of parable for what will be discussed more fully later.

Much of the progress in applied mathematics during the 20th century was due to understanding how to treat functions as vectors. This approach is especially helpful in solving linear differential equations in a general, complete way. We are familiar with the x-y plane. The x-y plane is a two-dimensional space and any point (x_0, y_0) in the plane can be represented as the vector (or the tip of the vector) $x_0 \hat{i} + y_0 \hat{j}$ where $\hat{i}$ and $\hat{j}$ are basis vectors parallel to the *x*- and *y*-axes, respectively. The x-y plane has two **dimensions** and is called **Euclidean two-space**. For example, if $\Re$ is the set of all real numbers, we can understand the point (4,3) as a point in the two-dimensional Euclidean space $\Re \times \Re$ or as the vector $4\hat{i} + 3\hat{j}$. The diagram below illustrates this:

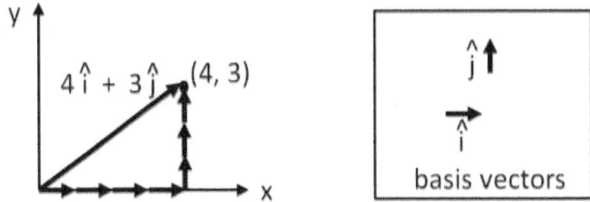

Representation using linearly independent vectors that are orthogonal.

Differential Equations in 24 Hours

In the same way that any point in Euclidean 2-space can be expressed in terms of $\hat{i}$ and $\hat{j}$, any solution to the differential equation $\ddot{u} + u = 0$ can be written as a linear combination of the **basis functions** $u_1 = \cos(t)$ and $u_2 = \sin(t)$. The set of all functions $u = u(t)$ that solve $\ddot{u} + u = 0$ is the **solution space** of $\ddot{u} + u = 0$. Any solution in the solution space of $\ddot{u} + u = 0$ can be represented as a linear combination $u = C_1 u_1 + C_2 u_2 = C_1 \cos(t) + C_2 \sin(t)$. The point (C_1, C_2) in the solution space of $\ddot{u} + u = 0$ plays the same role that the coordinate (x,y) played in the Euclidean two-space. It is a representation of any point in the space; thus it represents any solution of the differential equation. For example, the function $u = 4\cos(t) + 3\sin(t)$ is a solution of $\ddot{u} + u = 0$. In the solution space of $\ddot{u} + u = 0$, the solution function $4\cos(t) + 3\sin(t)$ can be identified with the point $(4,3)$ without ambiguity. Below is a diagram of the solution space:

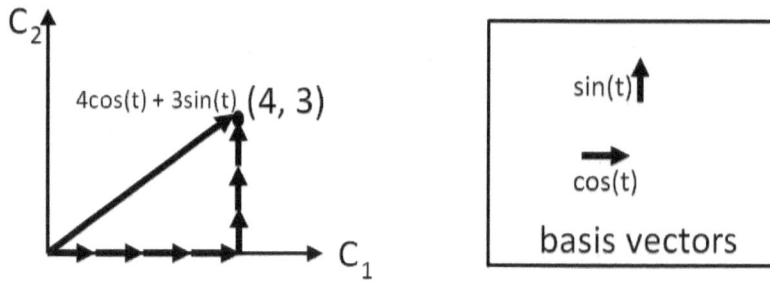

Representation using linearly independent basis functions that are orthogonal

We know that, in Euclidean two-space, we need two linearly independent vectors to serve as basis vectors. The strongest kind of linear independence occurs when the basis vectors are orthogonal; however, we could get by with using $\hat{e}_1 = \hat{i} + \frac{1}{2}\hat{j}$ and $\hat{e}_2 = \hat{j}$ since $\hat{e}_1$ and $\hat{e}_2$ in a linear combination can reach any point. For example, the point $(4,3)$ in the x-y plane is the tip of the arrow $4\hat{e}_1 + 1\hat{e}_2$. The vectors $\hat{e}_1$ and $\hat{e}_2$ are linearly independent: You cannot write one of these as a linear combination of the other. But linear independence is good enough to express any vector in the plane in the form $C_1\hat{e}_1 + C_2\hat{e}_2$.

Hour 12. Higher-order linear differential equations and their IVP's and BVP's

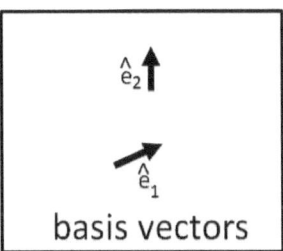

Representation using linearly independent vectors that are not orthogonal

What must be avoided, however, is **linear dependence** which occurs when one or more basis vectors can be expressed as a linear combination of one or more of the other basis vectors and there are only n vectors. For example, the point (4,3) in the x-y plane cannot be reached by:

$$C_1\hat{e}_1 + C_2\hat{e}_2 \text{ if } \hat{e}_1 = \hat{i} \text{ and } \hat{e}_2 = 2\hat{i} \qquad (12\text{-}1)$$

Since no combination of these can bring you above the x-axis.

Orthogonal vectors have the property that their inner product is equal to zero. This is true for basis vectors $\hat{i}$ and $\hat{j}$:

$$\hat{i} \cdot \hat{j} = 0 \qquad (12\text{-}2)$$

As we saw in Hour 11, the inner product between two functions g(t) and h(t) is formed by multiplying the first function times the conjugate of the second and integrating the result. If the inner product is zero then the functions are orthogonal.

For the differential equation $\ddot{u} + u = 0$, we can write it in the form $\ddot{u} = -u$ and ask: "What function(s) can I differentiate twice and obtain the negative of the original?" Sine has that property and so does cosine:

$$\frac{d^2}{dx^2}(\sin t) + \sin t = -\sin t + \sin t = 0$$

$$\frac{d^2}{dx^2}(\cos t) + \cos t = -\cos t + \cos t = 0 \tag{12-3}$$

But sin(t) and cos(t) are also orthogonal functions, since their inner product is zero; therefore, they represent a basis for the solution space of ü + u = 0. Any solution can be represented in the form:

$$u(t) = C_1 \cos(t) + C_2 \sin(t) \tag{12-4}$$

It is no coincidence that, for this 2nd-order differential equation, we need *two* linearly independent basis functions in order to express the general solution. The basis functions cos(t) and sin(t) are not only linearly independent; they are also orthogonal. We could represent the general solution in terms of two basis functions that are linearly independent but *not* orthogonal, say u(t) = $C_1 e_1(t)$ + $C_2 e_2(t)$ where $e_1(t)$ = cos(t) + ¼sin(t) and $e_2(t)$ = ¾sin(t). Even so, we will usually find it convenient, if possible, to find orthogonal basis functions.

We are ready now to talk about higher-order differential equations in greater generality. By *higher-order* we mean nth-order (also said "n-order") with *n* greater than one.

Definition 12-1: An nth-order linear ordinary differential equation, on the interval I, is an equation of the form:

$$a_n(x)\frac{d^n y}{dx^n} + a_{n-1}(x)\frac{d^{n-1} y}{dx^{n-1}} + \ldots + a_1(x)\frac{dy}{dx} + a_0(x)y = g(x)$$

$$\tag{12-5}$$

∎

Such equations are *linear* in the sense that the highest order of derivative of the dependent variable may be written as a linear function of any lower-order derivatives of the dependent variable:

Hour 12. Higher-order linear differential equations and their IVP's and BVP's

$$\frac{d^n y}{dx^n} = -\frac{a_{n-1}(x)}{a_n(x)}\frac{d^{n-1}y}{dx^{n-1}} - \frac{a_{n-2}(x)}{a_n(x)}\frac{d^{n-2}y}{dx^{n-2}} - \ldots - \frac{a_0(x)}{a_n(x)}y$$
$$+ \frac{g(x)}{a_n(x)}$$

(12-6)

To comprehend this, it helps to think of $-\dfrac{a_{n-1}(x)}{a_n(x)}$ as $m_1(x)$ and $-\dfrac{a_{n-2}(x)}{a_n(x)}$ as $m_2(x)$ in harkening back to the old "(dependent variable) = m(independent variable) + b" concept of "linear" from our age of innocence.

If you combine Eq. 12.5 with the initial conditions

$$y(x_0) = y_0, \quad y'(x_0) = y_1, \quad \ldots, \quad y^{(n-1)}(x_0) = y_{n-1} \quad (12\text{-}7)$$

Then you have a linear n^{th}-order initial-value problem. For such problems, the solution (on some interval I of the independent variable) is **unique** if all of the "a functions" as well as the function g are continuous and if:

$$a_n(x) \neq 0 \quad \text{for all } x \in I \quad (12\text{-}8)$$

Example 12-1: Solve the following IVP:

$$y'' + y = 0 \quad \text{with initial conditions } y(0) = 0 \text{ and } y'(0) = 1$$

(12-9)

The solution is going to be unique since $a_2(x)$, the leading "*a* function," does not equal zero and the other *a*'s and *g* are continuous. But first let's write down the general solution and then we'll determine the specific, unique solution to the given problem. As we have seen, sine and cosine are each solutions to the differential equation:

Differential Equations in 24 Hours

$$\frac{d^2}{dx^2}(\sin x) + \sin x = -\sin x + \sin x = 0$$

$$\frac{d^2}{dx^2}(\cos x) + \cos x = -\cos x + \cos x = 0$$

(12-10)

The linear combination $C_1 cos(x) + C_2 sin(x)$ also provides a solution:

$$\frac{d^2}{dx^2}(C_1 \cos x + C_2 \sin x) + C_1 \cos x + C_2 \sin x =$$

(12-11)

$$-C_1 \cos x + -C_2 \sin x + C_1 \cos x + C_2 \sin x = 0 \quad (12\text{-}12)$$

Indeed $C_1 cos(x) + C_2 sin(x)$ is the **general solution** to Eq. 12-9: it is a vector that reaches any point in the solution space of $y'' + y = 0$ and (C_1, C_2) can be used to represent any point in the solution space. So now we have only to apply the initial conditions and determine exactly which unique point (C_1, C_2) in the solution space is the ultimate solution to the problem. Taking

$$y = C_1 \cos(x) + C_2 \sin(x) \quad (12\text{-}13)$$

and plugging in the initial conditions, we have:

$$y(0) = C_1 \cos(0) + C_2 \sin(0) = C_1 = 0$$
$$y'(0) = -C_1 \sin(0) + C_2 \cos(0) = C_2 = 1$$

(12-14)

Thus we have been able to determine C_1 and C_2; therefore, the final unique solution to the present problem is:

$$y = \sin(x) \quad (12\text{-}15)$$

It is the point $(C_1, C_2) = (0,1)$ in the solution space with basis vectors $\hat{e}_1 = \cos(x)$, $\hat{e}_2 = \sin(x)$.

◊

Hour 12. Higher-order linear differential equations and their IVP's and BVP's

The harmonic oscillator example we have been using has the general solution with the auspicious vector space properties for more reasons than one. In addition to being a linear differential equation, it is also a homogeneous equation.

Definition 12-2: A **homogeneous linear differential equation** is one for which g(x) = 0 in Eq. 12-4:

$$a_n(x)\frac{d^n y}{dx^n} + a_{n-1}(x)\frac{d^{n-1} y}{dx^{n-1}} + \ldots + a_1(x)\frac{dy}{dx} + a_0(x)y = 0$$

(12-16)

■

This homogeneity of the equation has nothing to do with the "homogeneous coefficients" $M(x, y)$ and $N(x, y)$ that we discussed previously. Here homogeneous means that if y is a solution, then Cy is a solution. (You can look at the equation and see that C divides out.) Also, if y_1 and y_2 are solutions to the equation, then $C_1y_1 + C_2y_2$ is also a solution. That's why $C_1y_1 + C_2y_2$ is the "general solution" of the homogeneous ODE.

A solution space of a differential equation is the set of all possible solutions of that equation. A linear homogeneous ODE creates a **solution space** that has the same dimension as the order of the equation and the same number of basis functions as that order. The nth-order homogeneous linear ODE has its own n-dimensional solution space and n functions serving as basis functions. As we have alluded to already, there is a superposition principle at work when we write a solution as a linear combination of these n functions. If $y_1, y_2, \ldots y_k$ are solutions to:

$$a_k(x)\frac{d^k y}{dx^k} + a_{k-1}(x)\frac{d^{k-1} y}{dx^{k-1}} + \ldots + a_1(x)\frac{dy}{dx} + a_0(x)y = 0$$

(12-17)

Then

$$y = C_1 y_1 + C_2 y_2 + \ldots + C_k y_k$$

(12-18)

is also a solution.

81

Differential Equations in 24 Hours

If $y_1, y_2, \ldots y_n$ are linearly independent then they can be used as a **basis for the solution space** of the homogeneous equation:

$$a_n(x)\frac{d^n y}{dx^n} + a_{n-1}(x)\frac{d^{n-1} y}{dx^{n-1}} + \ldots + a_1(x)\frac{dy}{dx} + a_0(x)y = 0$$

(12-19)

Then the $y_1, y_2, \ldots y_n$ are linearly independent basis vectors of the space. The basis $\{y_1, y_2, \ldots y_n\}$ forms a coordinate system in which any solution $y = f(x)$ can be expressed:

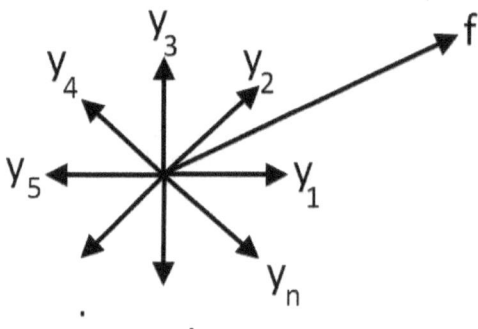

The basis vectors $\{y_1, y_2, \ldots, y_n\}$ can be tested for linear independence by calculating the Wronskian W, which is the following determinant:

$$W = \begin{vmatrix} y_1 & y_2 & \cdots & y_n \\ y_1' & y_2' & \cdots & y_n' \\ \vdots & \vdots & & \vdots \\ y_1'' & y_2'' & \cdots & y_n'' \end{vmatrix}$$

(12-20)

They are linearly independent if, for some point x_0 in I, W does not equal zero. Thus the condition for linear independence is:

Hour 12. Higher-order linear differential equations and their IVP's and BVP's

For $\{y_1, y_2, ..., y_n\}$ defined on I, if $W \neq 0$ for some $x_0 \in I$

then $\{y_1, y_2, ..., y_n\}$ are linearly independent.

(12-21)

Example 12-2: Show that the functions $\{\sin(x), \cos(x)\}$ are linearly independent by calculating the Wronskian. We have:

$$W = \begin{vmatrix} \sin(x) & \cos(x) \\ \cos(x) & -\sin(x) \end{vmatrix} = -\sin^2(x) - \cos^2(x) = -1 \neq 0$$

(12-22)

The Wronskian is not equal to zero; therefore, $\{\sin(x), \cos(x)\}$ are linearly independent.

◊

The quintessential example of a boundary value problem is an elastic string fixed at both ends. There are boundary conditions on the motion of the string (the fact that the walls are boundaries and so the string can't move up and down at the walls).

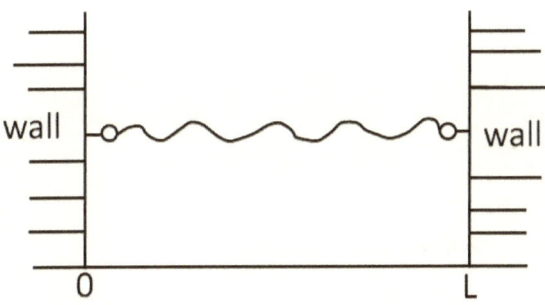

83

A **boundary value problem (BVP)** is a differential equation with boundary conditions. For the string, the walls provide boundaries which "contain" the motion, since a wave on the string cannot continue into the wall (except by a very small amount). The physical system above can be modeled using the following BVP:

$$y'' + \omega_m^2 y = 0$$

With boundary conditions y(0) = 0, y(L) = 0. (12-23)

The subscript m appearing on the parameter omega is to indicate that there will be harmonics. (Just like a guitar string). This is really a whole bunch of different equations…one for every different value of m. The general solution is:

$$y = C_1 \cos(\omega_m x) + C_2 \sin(\omega_m x)$$

(12-24)

When we plug in the first boundary condition we obtain:

$$y(0) = C_1 = 0$$

(12-25)

which kills off the first term right of the equals sign in equation Eq. 12-24. Plugging in the second boundary condition, we see that:

$$y(L) = C_2 \sin(\omega_m L) = 0,$$

which requires that $\omega_m L = q\pi$, $q = ...,-2,-1,0,1,2,...$

(12-26)

So the solutions are:

$$y(x) = C \sin(\frac{q\pi}{L} x)$$

(12-27)

The solutions look like this for some different values of q:

84

Hour 12. Higher-order linear differential equations and their IVP's and BVP's

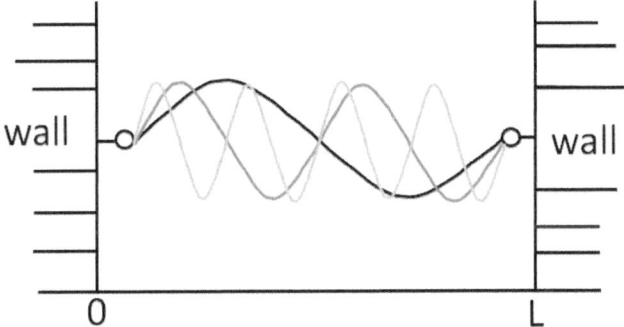

So far we've talked about higher-order linear ODE's that are homogeneous (meaning g(x) = 0). What we've learned so far about homogeneous equations is a great leg-up on solving the non-homogeneous equation:

$$a_n(x)\frac{d^n y}{dx^n} + a_{n-1}(x)\frac{d^{n-1} y}{dx^{n-1}} + \ldots + a_1(x)\frac{dy}{dx} + a_0(x) y = g(x)$$

(12-28)

In order to solve the non-homogeneous equation, we first find the general solution $C_1 y_1 + C_2 y_2 + \ldots + C_n y_n$ to the homogeneous equation formed by setting g(x) = 0. (This will be called the homogeneous solution y_h.) Then we find one particular solution y_p to the *full* non-homogeneous solution and add it to the homogeneous solution y_h to form the general solution y_g: The general solution to the non-homogeneous equation then is:

$$y_g = y_h + y_p$$

(12-29)

Or:

$$y_g = C_1 y_1 + C_2 y_2 + \ldots + C_n y_n + y_p$$

(12-30)

Thus an algorithm for solving the non-homogeneous linear ODE, Eq. 12-28, is:

1.) Set g(x) = 0, solve the homogeneous equation and establish the homogeneous part of the solution $y_h = C_1 y_1 + \ldots + C_n y_n$.

85

Differential Equations in 24 Hours

2.) Find a particular solution y_p for the non-homogeneous case.

3.) Then the general solution of the overall non-homogeneous problem is $y_g = y_h + y_p$.

Let's apply this algorithm to an example. At the same time, we will try out a new technique for solving homogeneous linear ODE's with constant coefficients. For such equations, we look for solutions in the form $y = e^{ax}$. We will explore this method more fully in the next Hour.

Example 12-3: Solve:

$$y'' - 4y = 6x - 4x^3 \qquad (12\text{-}31)$$

Solution: The first step is to set $g(x) = 0$ and solve the corresponding homogeneous equation:

$$y'' - 4y = 0 \qquad (12\text{-}32)$$

I need two independent solutions since $y'' - 4y = 0$ is 2nd order. (These two will form the basis for the solution space of the homogeneous equation.) Look for solutions in the form $y = e^{ax}$. The second derivative is $y'' = a^2 e^{ax}$. Plugging this in we get:

$$y'' - 4y = a^2 e^{ax} - 4e^{ax} = 0 \qquad (12\text{-}33)$$

Dividing out e^{ax}, we obtain the **auxiliary** equation:

$$a^2 - 4 = 0 \qquad (12\text{-}34)$$

Which gives us two values for a:

$$a = \pm 2 \qquad (12\text{-}35)$$

Since, by looking for solutions in the form $y = e^{ax}$, we came up with $a = \pm 2$, we have two solutions to the homogeneous equation: $y_1 = e^{2x}$ and $y_2 = e^{-2x}$. We can show that these are independent by demonstrating a non-zero Wronskian:

Hour 12. Higher-order linear differential equations and their IVP's and BVP's

$$W = \begin{vmatrix} e^{2x} & e^{-2x} \\ 2e^{2x} & -2e^{-2x} \end{vmatrix} = -2 - 2 = -4 \neq 0 \quad (12\text{-}36)$$

So the homogeneous solution y_h is:

$$y_h = C_1 e^{2x} + C_2 e^{-2x} \quad (12\text{-}37)$$

Now we have only to find one single particular solution y_p, then we can add this to the homogeneous solution and finish solving the non-homogeneous differential equation. For the time being, we will find y_p by guessing. Let's try $y_p = x^3$ since it's one of the terms in g(x). Plugging in we see that:

$$y_p'' - 4y_p = 6x - 4x^3 \quad (12\text{-}38)$$

Thus the non-homogeneous equation is solved by this particular solution. (We will soon learn how to determine particular solutions without guessing.) The answer to the original problem then is the general solution $y_g = y_h + y_p$ which is the sum of the homogeneous and the particular solution:

$$y = y_h + y_p = C_1 e^{2x} + C_2 e^{-2x} + x^3 \quad (12\text{-}39)$$

Here no boundary or initial conditions are presented in the problem statement; therefore, we are done.

◊

The proof that $y_h + y_p$ is a solution to Eq. 12-28 is one of the easiest proofs in differential equations. If y_h is the homogeneous solution, then:

$$a_n(x)\frac{d^n y_h}{dx^n} + \ldots + a_1(x)\frac{dy_h}{dx} + a_0(x)y_h = 0$$

$$(12\text{-}40)$$

87

Differential Equations in 24 Hours

If y_p is a particular solution, then

$$a_n(x)\frac{d^n y_p}{dx^n} + \ldots + a_1(x)\frac{dy_p}{dx} + a_0(x)y_p = g(x)$$

(12-41)

If we add Eq. 12-39 and Eq. 12-40 together, we obtain:

$$a_n(x)\frac{d^n(y_h+y_p)}{dx^n} + \ldots + a_1(x)\frac{d(y_h+y_p)}{dx} + a_0(x)(y_h+y_p) = g(x)$$

(12-42)

So we have proved that $y_h + y_p$ must also be a solution.

EXERCISES

Exercise 12-1:

Three solutions of $\ddot{y} + \omega^2 y = 0$ are:

$$\left\{\sin(\omega t),\ \cos(\omega t),\ \sqrt{2}\cos\left(\omega t - \frac{\pi}{4}\right)\right\}.$$

Show that these are not linearly independent by writing the third function as a linear combination of the first two.

Hour 12. Higher-order linear differential equations and their IVP's and BVP's

Exercise 12-2: Calculate the Wronskian for the functions of the previous exercise:

$$W = \begin{vmatrix} y_1 & y_2 & y_3 \\ y_1' & y_2' & y_3' \\ y_1'' & y_2'' & y_3'' \end{vmatrix}$$

(12-43)

What is your conclusion?

Exercise 12-3: A very simplified radio consists of an inductor L and a capacitor C. It is picking up a weak radio signal $\sin(\omega_0 t)$ at frequency ω_0.

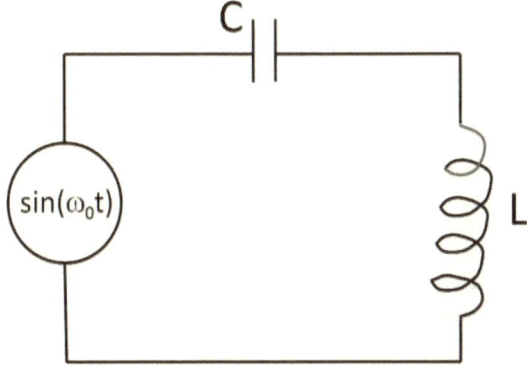

The current j(t) is the rate at which charges Q are flowing per unit time. The voltage drop across the inductor is L(dj/dt) and the voltage drop across the capacitor is Q/C. Kirchoff's Law says that these voltage drops must add up to the applied voltage:

$$L\frac{dj}{dt} + \frac{1}{C}Q = \sin(\omega_0 t)$$

(12-44)

The current j is the time derivative of charge: j(t) = dQ/dt. Differentiating Eq. 12-44 we obtain a second order differential equation:

$$L\frac{d^2 j}{dt^2} + \frac{1}{C}j = \omega_0 \cos(\omega_0 t)$$

(12-45)

Differential Equations in 24 Hours

Which can be placed in the form of Eq. 8-2 by dividing through by L:

$$\frac{d^2 j}{dt^2} + \frac{1}{LC} j = \frac{\omega_0}{L} \cos(\omega_0 t) \tag{12-46}$$

Show that the homogeneous solution is:

$$j_h = C_1 \cos(\frac{1}{\sqrt{LC}} t) + C_2 \sin(\frac{1}{\sqrt{LC}} t) \tag{12-47}$$

Exercise 12-4: Show that a particular solution to the differential equation of the previous problem is:

$$j_p = \frac{\omega_0 C}{1 - \omega_0^2 LC} \cos(\omega_0 t) \tag{12-48}$$

Exercise 12-5: Write down the general solution to Eq. 12-45.

Exercise 12-6: To make a more realistic radio, we can add a resistor R.

$$L\frac{d^2 j}{dt^2} + R\frac{dj}{dt} + \frac{1}{C} j = \omega_0 \cos(\omega_0 t) \tag{12-49}$$

A particular solution to this is:

$$j_p = \frac{1}{\sqrt{R^2 + \left[\frac{1}{\omega_0 C} - \omega_0 L\right]^2}} \cos(\omega_0 t) \tag{12-50}$$

Note that this reduces to Eq. 12-46 when R = 0. Suppose the radio is now made tunable by providing for the capacitor to have adjustable capacitance C. What value of capacitance C provides the largest amplitude for y_p? (This electrical resonance is the method by which radios tune to channels.)

Hour 12. Higher-order linear differential equations and their IVP's and BVP's

Exercise 12-7: A guitarist plucks a guitar string which has a length of L between places where it is tightly fixed at $x = 0$ and $x = L$.

Assuming the same boundary value problem as Eq. 12-23, what is the lowest non-zero frequency that can be heard?

Exercise 12-8: If the guitarist places his finger at $x = L/2$, and plucks to the right of that, what is the lowest non-zero frequency that can be heard.

Hour 13. Homogeneous linear ODE's with constant coefficients

A homogeneous linear equation with constant coefficients is an equation in the form:

$$a_n y^{(n)} + a_{n-1} y^{(n-1)} + \ldots + a_1 y' + a_0 y = 0 \qquad (13\text{-}1)$$

Where all the a's are constants. All solutions are either exponentials or they are made up of exponentials, possibly complex exponentials. So the approach here is to look for solutions of the form $y = e^{mx}$. This results in a polynomial equation in m which we then try to solve for the roots.

Example 13-1:   Solve:

$$y'' - 5y' + 6y = 0 \qquad (13\text{-}2)$$

Letting $y = e^{mx}$ we obtain $y' = me^{mx}$ and $y'' = m^2 e^{mx}$

$$(13\text{-}3)$$

Plugging in, we have:

$$y'' - 5y' + 6y = m^2 e^{mx} - 5me^{mx} + 6e^{mx} = 0$$

$$(13\text{-}4)$$

We can divide out e^{mx} and what we're left with is called the "characteristic equation" or the "auxiliary equation."

$$m^2 - 5m + 6 = 0 \qquad (13\text{-}5)$$

This can be factored as $(m - 2)(m - 3) = 0$ or $m_1 = 2$ and $m_2 = 3$. So, by assuming $y = e^{mx}$ we have arrived at two solutions e^{2x} and e^{3x}. The general solution then is:

$$y = C_1 e^{2x} + C_2 e^{3x} \qquad (13\text{-}6)$$

◊

Hour 13. Homogeneous linear ODE's with constant coefficients

When using this $y = e^{mx}$ trick for solving second-order equations, that is, equations of the form

$$ay'' + by' + cy = 0 \qquad (13\text{-}7)$$

There are three things that can happen:

1.) if $b^2 - 4ac > 0$ then m_1 and m_2 are real and distinct.

2.) if $b^2 - 4ac = 0$ then $m_1 = m_2$ (we have a double root).

3.) if $b^2 - 4ac < 0$ then m_1 and m_2 are complex conjugate (meaning that if, say, $m_1 = 2i$, then $m_2 = -2i$.

Case one is the most simple; we obtain two distinct m's; therefore, this gives us two independent functions and we have what we need to write the general solution. Case two gives me only one solution; therefore, we have to use some other technique (for instance **reduction of order**) to get the other solution. (We will discuss reduction of order more fully later on. One means by which to implement reduction of order involves looking for a second solution $y_2(x)$ as the product of the first solution $y_1(x)$ and some function $u(x)$. (That is, plug in $y_2(x) = u(x)y_1(x)$.)) Case three might be reducible to real using Euler's formula:

$$e^{ix} = \cos(x) + i\sin(x) \qquad (13\text{-}8)$$

Example 13-1 was an example of Case 1: The auxiliary equation has two distinct, real roots. Now let's look at a Case 2 example, an example in which the auxiliary equation has a double root:

Example 13-2: Case 2: A double root. Solve:

$$y'' - 6y' + 9y = 0 \qquad (13\text{-}9)$$

Our trial solution is $y = e^{mx}$. The derivatives of this are $y' = me^{mx}$ and $y'' = m^2 e^{mx}$. Plugging these into Eq. 13-9, we obtain:

$$m^2 e^{mx} - 6me^{mx} + 9e^{mx} = 0 \qquad (13\text{-}10)$$

Differential Equations in 24 Hours

Which gives the characteristic equation:

$$m^2 - 6m + 9 = 0 \quad \text{which factors as} \quad (m - 3)^2 = 0 \tag{13-11}$$

This gives the double root m = 3. So we only obtain one basis function: $y_1 = e^{3x}$. But we need two basis functions y_1 *and* y_2 to express the general solution of a second-order linear ODE. Now use reduction of order to obtain the second solution y_2. We look for a second solution in the form $y_2 = u(x)y_1$. Here we seek a solution in the form:

$$y_2 = u(x)e^{3x} \tag{13-12}$$

Then:

$$\begin{aligned} y_2' &= u'e^{3x} + 3ue^{3x} \\ y_2'' &= u''e^{3x} + 3u'e^{3x} + 3u'e^{3x} + 9ue^{3x} \end{aligned} \tag{13-13}$$

So, plugging into Eq. 13-9, we get:

$$u''e^{3x} + 6u'e^{3x} + 9ue^{3x} - 6u'e^{3x} - 18ue^{3x} + 9ue^{3x} = 0 \tag{13-14}$$

Lucky for us, all of this collapses down to a succinct little equation:

$$u''e^{3x} = 0 \quad \text{or:} \quad u'' = 0 \tag{13-15}$$

Integrating, we are able to find a second solution:

$$u = C_1 x + C_2 \tag{13-16}$$

Thus:

$$y = ue^{3x} = C_1 x e^{3x} + C_2 e^{3x} \tag{13-17}$$

Hour 13. Homogeneous linear ODE's with constant coefficients

We can show that the basis functions $\{xe^{3x}, e^{3x}\}$ are linearly independent by verifying that the Wronskian is not zero:

$$W = \begin{vmatrix} xe^{3x} & e^{3x} \\ (e^{3x} + 3xe^{3x}) & 3e^{3x} \end{vmatrix} = 3xe^{6x} - e^{6x} - 3xe^{6x} = -e^{6x}$$

(13-18)

There is no real x that will make $-e^{6x}$ equal to zero; therefore $\{xe^{3x}, e^{3x}\}$ is a basis for the solution space of Eq. 13-9 and Eq. 13-17 is the general solution.

◊

Example 13-3: Case 3: Complex conjugate roots. Solve the harmonic oscillator:

$$y'' + y = 0 \qquad (13\text{-}19)$$

Again, our trial solution is $y = e^{mx}$. The second derivative of this is $y'' = m^2 e^{mx}$. Plugging these in, we obtain:

$$m^2 e^{mx} + 1e^{mx} = 0 \quad \text{which gives} \quad m^2 + 1 = 0$$

(13-20)

The solution to the auxiliary equation is a couple of complex conjugate roots:

$$m = \pm i \qquad (13\text{-}21)$$

Thus, the solution this gives us is:

$$y = C_1 e^{ix} + C_2 e^{-ix} \qquad (13\text{-}22)$$

Using Euler's formula we have:

$$y = C_1(\cos(x) + i\sin(x)) + C_2(\cos(x) - i\sin(x))$$
$$y = (C_1 + C_2)\cos x + i(C_1 - C_2)\sin x$$

(13-23)

We are only interested in solutions to this which are real. Taking the real component of this, we have $y_1 = \cos(x)$. We can use reduction of order to get another linearly independent solution y_2 in order to be able to write the general solution. Let $y_2 = u(x)\cos(x)$. The second derivative of this is $y_2'' = u''\cos(x) - 2u'\sin(x) - u\cos(x)$. Plugging into Eq. 13-19, we have:

$$u''\cos(x) - 2u'\sin(x) - u(x)\cos(x) + u(x)\cos(x) = 0$$

(13-24)

Notice that the terms involving $u(x)$ cancel each other. Now define $w(x) = u'$ in order to reduce order to first-order:

$$w'\cos(x) - 2\sin(x)w = 0 \qquad (13\text{-}25)$$

We can use separable variables to solve for w:

$$\frac{dw}{w} - 2\tan(x)\,dx = 0$$

(13-26)

Integrating, we have:

$$\ln(w) + 2\ln(\cos(x)) = \ln C \qquad (13\text{-}27)$$

Which may be written as:

$$w = C_1 \sec^2 x = u' \qquad (13\text{-}28)$$

Integrating, we determine u:

$$u(x) = C_1 \tan x + C_2 \qquad (13\text{-}29)$$

We said that the second solution is the product of $u(x)$ and the first solution $\cos(x)$:

Hour 13. Homogeneous linear ODE's with constant coefficients

$$y_2(x) = u(x)y_1(x)$$
$$= (C_1 \tan x + C_2)\cos x = C_1 \sin x + C_2 \cos x$$

(13-30)

As we've shown before, {sinx, cosx} is a linearly independent set of basis functions for the solution space; therefore, we have what we need to present the general solution:

$$y(x) = C_1 \sin x + C_2 \cos x \qquad (13\text{-}31)$$

◊

As the previous example showed, the case of complex conjugate roots can be a chore. To simplify matters, we have the following theorem:

Theorem:

If the auxiliary equation has distinct complex conjugate roots $m = \alpha \pm \beta i$, then the general solution to $y'' + by' + cy = 0$ is:

$$y = e^{\alpha x}(C_1 \cos(\beta x) + C_2 \sin(\beta x)) \qquad (13\text{-}32)$$

•

Example 13-4: Solve:

$$y'' - 14y' + 74y = 0 \qquad (13\text{-}33)$$

Plugging in $y = e^{mx}$ and the derivatives $y' = me^{mx}$ and $y'' = m^2 e^{mx}$, we have, after dividing out the common term e^{mx}:

$$m^2 - 14m + 74 = 0 \qquad (13\text{-}34)$$

The quadratic formula gives us:

$$m = 7 \pm i5 = \alpha \pm i\beta \qquad (13\text{-}35)$$

Differential Equations in 24 Hours

Thus the general solution is:

$$y = e^{7x}(C_1 \cos(5x) + C_2 \sin(5x)) \qquad (13\text{-}36)$$

◊

Be careful to keep in mind that the above theorem only applies to constant coefficient, second-order linear ODE's.

We can summarize the situation for a 2nd-order linear ODE with constant coefficients in a convenient table as follows.

Table 13-1: *Homogeneous solutions for 2nd-order linear ODE with constant coefficients.*

Case		
Case 1	Distinct real roots	$y_h = C_1 e^{m_1 x} + C_2 e^{m_2 x}$
Case 2	Repeated real root	$y_h = C_1 e^{mx} + C_2 x e^{mx}$
Case 3	Complex roots	$y_h = e^{\alpha x}[C_1 \cos(\beta x) + C_2 \sin(\beta x)]$

EXERCISES

Exercise 13-1: Find the general solution of:

$$\frac{d^2 y}{dx^2} - 3\frac{dy}{dx} + 2y = 0 \qquad (13\text{-}37)$$

Exercise 13-2: Find the general solution of:

$$\ddot{y} - 2\dot{y} + 5y = 0 \qquad (13\text{-}38)$$

Hour 13. Homogeneous linear ODE's with constant coefficients

Exercise 13-3: Find the general solution of:

$$y'' - 8y' + 16y = 0 \qquad (13\text{-}39)$$

Exercise 13-4: Solve the IVP:

$$y'' - (2 + \pi)y' + 2\pi y = 0$$

With initial conditions $y(0) = 1$ and $y'(0) = \pi$.

$$(13\text{-}40)$$

Hour 14. Reduction of order

In the previous Hour, we used reduction of order to find a second solution to a differential equation whose auxiliary equation had a double root. This was accomplished by making the substitution $y_2(x) = u(x)y_1(x)$, a method that Euler originated. It turns out that there are several techniques that accomplish reduction of order involving several different types of substitutions. The problem at hand determines which approach to use.

The simplest reduction of order technique involves making the substitution $u(x) = y'$. An obvious occasion to use this is when y does not appear explicitly in a differential equation:

Reduction of order algorithm 1: Consider the 2nd-Order Linear DE:

$$a_2(x)y'' + a_1(x)y' + a_0(x)y = b(x) \qquad (14\text{-}1)$$

If $a_0(x) = 0$ then we can reduce the order by simply setting:

$$u = y' \qquad (14\text{-}2)$$

•

This substitution reduces the equation to

$$a_2(x)u' + a_1(x)u = b(x) \qquad (14\text{-}3)$$

which is linear, first-order and can be solved using techniques we know already.

Example 14-1: Solve:

$$xy'' + y' = 0 \qquad (14\text{-}4)$$

Solution:

$$u = y' \quad \text{and} \quad u' = y'' \qquad (14\text{-}5)$$

Thus the equation reduces to:

$$xu' + u = 0 \qquad (14\text{-}6)$$

100

Hour 14. Reduction of order

Or:

$$x\frac{du}{dx} = -u \tag{14-7}$$

$$\frac{du}{u} = -\frac{dx}{x} \tag{14-8}$$

This integrates to become:

$$\ln u = -\ln x + \ln C \tag{14-9}$$

$$u = \frac{C}{x} \tag{14-10}$$

Substituting u = y' back in, we have:

$$y' = \frac{C}{x} \tag{14-11}$$

Thus the general solution is:

$$y = C_1 \ln|x| + C_2 \tag{14-12}$$

◊

A method also called reduction of order, which we have some experience with from Hour 13, involves making the substitution:

$$y_2(x) = u(x)y_1(x) \tag{14-13}$$

To reduce the order. This is a handy tool when we already have one solution $y_1(x)$ and we are seeking a second solution $y_2(x)$

Reduction of order algorithm 2:

If we have:

Differential Equations in 24 Hours

$$a_2(x)y'' + a_1(x)y' + a_0(x)y = 0 \qquad (14\text{-}14)$$

And if we have a non-trivial solution $y_1(x)$, then we can find a second, linearly independent solution $y_2(x)$ by substituting:

$$y_2(x) = u(x)y_1(x) \qquad (14\text{-}15)$$

This substitution causes the 2nd-order DE to reduce to first order.

•

Example 14-2: Find the general solution to the following differential equation, given one solution $y_1(x)$:

$$y'' - 4y' + 4y = 0 \qquad \text{has solution} \quad y_1 = e^{2x}$$

$$(14\text{-}16)$$

Solution:

We form $y_2 = uy_1$ and differentiate:

$$y_2' = u'y_1 + uy_1' = u'e^{2x} + u2e^{2x} \qquad (14\text{-}17)$$

Differentiating again, we get:

$$y_2'' = e^{2x}u'' + 4e^{2x}u' + 4e^{2x}u \qquad (14\text{-}18)$$

Plugging y_2, y_2', and y_2'' into the differential equation we obtain:

$$e^{2x}u'' + 4e^{2x}u' + 4e^{2x}u - 4u'e^{2x} - 8ue^{2x} + 4ue^{2x} = 0$$

$$(14\text{-}19)$$

102

Hour 14. Reduction of order

Which reduces to:

$$u'' = 0 \tag{14-20}$$

Integrating twice we obtain:

$$u = C_1 x + C_2 \tag{14-21}$$

Thus the second solution we are looking for is:

$$y_2 = u y_1 = C_1 e^{2x} x + C_2 e^{2x} \tag{14-22}$$

The functions $\{xe^{2x}, e^{2x}\}$ are linearly independent:

$$W = \begin{vmatrix} xe^{2x} & e^{2x} \\ e^{2x}(1+2x) & 2e^{2x} \end{vmatrix} = 2xe^{4x} - (1+2x)e^{4x} = e^{4x} \tag{14-23}$$

This is owing to the fact that the Wronskian is not zero. That being the case, the general solution is:

$$y = C_1 x e^{2x} + C_2 e^{2x} \tag{14-24}$$

◊

Let us understand the linear algebra in the second algorithm. We have a 2nd order linear DE here so we can expect the solution space to be spanned by two solutions y_1 and y_2 that are independent. We are given y_1. What does it mean to look for the other solution to be $y_2 = u(x)y_1$? This is a very important point. The answer is that, by setting $y_2 = u(x)y_1$, we are coercing the answer to not be $y_2 = Cy_1$, which is a simple multiple of y_1. Again, by forcing the two solutions not to be multiples of each other we assure they are not simple multiples of each other.

There is one other method known as reduction of order which is important to mention. This occurs when the independent variable x does not explicitly appear. In such cases, a subtle chain rule trick may be used to reduce order.

Differential Equations in 24 Hours

<u>Reduction of order algorithm 3:</u>

If we have the constant coefficient 2nd-order equation:

$$a_2 y'' + a_1 y' + a_0 y = b \qquad (14\text{-}25)$$

Then, defining:

$$u = \frac{dy}{dx} \qquad (14\text{-}26)$$

We note that:

$$y'' = u' = \frac{du}{dx} = \frac{du}{dy}\frac{dy}{dx} = u\frac{du}{dy} \qquad (14\text{-}27)$$

Replacing y' and y'' using these last two results, Eq. 14-25 becomes first-order:

$$a_2 u \frac{du}{dy} + a_1 u + a_0 y = b \qquad (14\text{-}28)$$

•

Example 14-3: Use reduction of order to solve $y'' + y = 0$.

Solution: Making the substitution for y'' provided by Eq. 14-27, we have:

$$u\frac{du}{dy} + y = 0 \qquad (14\text{-}29)$$

Multiplying through by dy and integrating, we have:

$$\frac{u^2}{2} + \frac{y^2}{2} = \frac{C_1}{2} \qquad (14\text{-}30)$$

Hour 14. Reduction of order

Thus we can solve for u and re-substitute:

$$u = \frac{dy}{dx} = \pm\sqrt{C_1 - y^2} \qquad (14\text{-}31)$$

Separating variables, and factoring out C_1 from the radical, we obtain:

$$\frac{dy}{\sqrt{C_1 - y^2}} = \pm dx \qquad (14\text{-}32)$$

The left side integrates to become the arcsine:

$$\arcsin\frac{y}{\sqrt{C_1}} = C_2 \pm x \qquad (14\text{-}33)$$

Taking the sine of each side, and multiplying by the constant, we have, after reducing the sine of a sum:

$$y = \left(\sqrt{C_1}\sin(C_2)\right)\cos(x) \pm \left(\sqrt{C_1}\cos(C_2)\right)\sin(x) \qquad (14\text{-}34)$$

The coefficients $\sqrt{C_1}\sin(C_2)$ and $\sqrt{C_1}\cos(C_2)$ may be used to represent any two real numbers; therefore, we are justified in presenting the solution as:

$$y = C_1\cos(x) + C_2\sin(x) \qquad (14\text{-}35)$$

◊

EXERCISES

Are linear combinations $C_1 y_1 + C_2 y_2$ also solutions to $y'' + y'y = 0$? If not, why not?

Exercise 14-1: Find a particular solution to:

$$xy'' + 2y' = x^2 \qquad (14\text{-}36)$$

Differential Equations in 24 Hours

Exercise 14-2: Find the general solution to:

$$y'' - y' + \frac{1}{4}y = 0 \tag{14-37}$$

Exercise 14-3: For the non-linear second-order equation $y'' + y'y = 0$, find two particular solutions y_1 and y_2 by using the substitutions of the third algorithm:

$$y' = u \quad \text{and} \quad y'' = u\frac{du}{dy} \tag{14-38}$$

Exercise 14-4: For $x^2 y'' - xy' + y = 0$ one solution is: $y_1(x) = x$. Use reduction of order to find a second solution.

Hour 15. Undetermined coefficients

The non-homogeneous linear ODE with constant coefficients is:

$$a_n \frac{d^n y}{dx^n} + a_{n-1} \frac{d^{n-1} y}{dx^{n-1}} + \ldots + a_1 \frac{dy}{dx} + a_0 y = g(x)$$

(15-1)

For this equation, our ultimate objective is to find the general solution $y_g = y_h + y_p$. We already know how to find a homogeneous solution y_h to the corresponding homogeneous equation $a_n y^{(n)} + a_{n-1} y^{(n-1)} + \ldots + a_1 y' + a_0 y = 0$. The homogeneous solution is a linear combination of linearly independent basis vectors $\{y_1, y_2, \ldots, y_n\}$. The method of undetermined coefficients is helpful for finding a particular solution y_p to the *non-homogeneous* case.

In order to apply the method of undetermined coefficients, $g(x)$ must consist of polynomials, exponentials, sines, and cosines in linear combinations. The reason for this is that these functions are "closed under differentiation." If you differentiate sines and cosines, you get sines and cosines (up to a multiplicative constant). The same is true for exponentials and polynomials.

The undetermined coefficients algorithm (superposition approach) goes as follows. There are five steps:

<u>Undetermined coefficients algorithm:</u>

1.) Solve the homogeneous equation to obtain the set S_h of homogeneous solutions:

$$S_h = \{y_1, y_2, \ldots, y_n\}$$

(15-2)

2.) Next, write down each term of $g(x)$ together with all of its derivatives, ignoring multiplicative constants.

3.) Write down a temporary y_{p_temp} as a linear combination of these terms, grouping them by family using parentheses (polynomials together, exponentials together, and sines and cosines together).

4.) Compare the components of y_{p_temp} with the elements of S_h. If there is a duplication within a family, multiply the family term by the lowest power of x

Differential Equations in 24 Hours

that will remove all such duplication to form y_p.

5.) Plug y_p into 15-1 and solve for the undetermined coefficients.

Example 15-1: Solve the constant coefficient non-homogeneous equation:

$$y^{(4)} - y'' = x^2 - e^{2x} \tag{15-3}$$

Solution:

Step 1.) The first step is to solve the homogeneous equation:

$$y^{(4)} - y'' = 0 \tag{15-4}$$

Any solution y = A, a constant, and y = Bx, where B is a constant, are solutions to the homogeneous equation. Exponentials will also be solutions. Letting $y = e^{mx}$, we find the characteristic equation:

$$y'' = m^2 e^{mx}, \quad y^{(4)} = m^4 e^{mx} \tag{15-5}$$

Plugging in,

$$m^4 e^{mx} - m^2 e^{mx} = 0 \tag{15-6}$$

Which becomes:

$$m^2 - 1 = 0 \tag{15-7}$$

So e^{-x} and e^x are also solutions of the homogeneous equation. Thus the set of homogeneous solutions is:

$$S_h = \left\{1, x, e^{-x}, e^x\right\} \tag{15-8}$$

Hour 15. Undetermined coefficients

Step 2.) Write down each term of g(x) together with all of its derivatives:

$$1, x, x^2, e^{2x} \qquad (15\text{-}9)$$

Step 3.) Write down a candidate y_{p_temp} as a linear combination of these terms, grouping them by family using parentheses:

$$y_{p\_temp} = (Cx^2 + Dx + E) + (Fe^{2x}) \qquad (15\text{-}10)$$

4.) Compare the components of y_{p_temp} with the elements of S_h. If there is a duplication within a family, multiply the family term by the lowest power of x that will remove all such duplication to form y_p. This requires multiplying the polynomial family by x^2:

$$y_p = x^2(Cx^2 + Dx + E) + (Fe^{2x}) \qquad (15\text{-}11)$$

5.) Plug y_p into 15-3 and solve for the undetermined coefficients.

$$\begin{aligned}
y_p &= Cx^4 + Dx^3 + Ex^2 + Fe^{2x} \\
y_p' &= 4Cx^3 + 3Dx^2 + 2Ex + 2Fe^{2x} \\
y_p'' &= 12Cx^2 + 6Dx + 2E + 4Fe^{2x} \\
y_p''' &= 24Cx + 6D + 8Fe^{2x} \\
y_p^{(4)} &= 24C + 16Fe^{2x}
\end{aligned}$$

$$(15\text{-}12)$$

Plugging in, we have:

$$(24C - 2E) - 6Dx - 12Cx^2 + 12Fe^{2x} = x^2 - e^{2x} \qquad (15\text{-}13)$$

Matching terms on the left- and right-hand sides of the equation, we have:

$$\left.\begin{array}{r}24C - 2E = 0 \\ D = 0 \\ -12C = 1 \\ 12F = -1\end{array}\right\} \Rightarrow \begin{array}{l}C = -1/12 \\ D = 0 \\ E = -1 \\ F = -1/12\end{array}$$

(15-14)

Thus:

$$y_p = -\frac{1}{12}x^4 - x^2 - \frac{1}{12}e^{2x} \qquad (15\text{-}15)$$

And:

$$y_g = C_1 + C_2 x + C_3 e^{-x} + C_4 e^x - \frac{1}{12}x^4 - x^2 - \frac{1}{12}e^{2x} \quad (15\text{-}16)$$

◊

Example 15-2: Solve:

$$y'' + y = 2\cos x \qquad (15\text{-}17)$$

Solution:

Step 1.) The set of functions that solve the homogeneous equation $y'' + y = 0$ are $S_h = \{\sin x, \cos x\}$.

Step 2.) We write down each term of $g(x)$ together with all of its derivatives (ignoring constants):

$$\sin x, \cos x \qquad (15\text{-}18)$$

Step 3.) Writing down a candidate y_{p_temp} as a linear combination of these terms, grouping them by family using parentheses:

$$y_{p\_temp} = A\sin x + B\cos x \qquad (15\text{-}19)$$

110

Hour 15. Undetermined coefficients

4.) Comparing the components of y_{p_temp} with the elements of S_h we see there is duplication which can be removed by multiplying by x; therefore, we do so.

$$y_p = Ax\sin x + Bx\cos x \qquad (15\text{-}20)$$

5.) We plug this into 15-17 and solve for the undetermined coefficients:

$$\begin{aligned} y'_p &= A\sin x + Ax\cos x + B\cos x - Bx\sin x \\ y''_p &= 2A\cos x - 2B\sin x - Ax\sin x - Bx\cos x \end{aligned} \qquad (15\text{-}21)$$

Plugging in,

$$2A\cos x - 2B\sin x - Ax\sin x - Bx\cos x + Ax\sin x + Bx\cos x = 2\cos x \qquad (15\text{-}22)$$

Which becomes:

$$2A\cos x - 2B\sin x = 2\cos x \qquad (15\text{-}23)$$

Thus B = 0 and A = 1. Therefore:

$$\begin{aligned} y_p &= x\sin x \\ y_g &= A\sin x + B\cos x + x\sin x \end{aligned} \qquad (15\text{-}24)$$

◊

EXERCISES

Exercise 15-1: Find the general solution to $y'' + 3y' - 10y = 6e^{4x}$.

Exercise 15-2: Find the general solution to $y'' + 4y = 3\sin(x)$.

Exercise 15-3: Find the general solution to $2y'' + 3y' + y = \cos(x)$.

Exercise 15-4: Find the general solution to $y'' - 2y' - 3y = 2e^{3x} + 8xe^{3x}$.

Hour 16. Variation of parameters

Wronski was born in Poland but spent most of his life in France. We have been introduced to the Wronskian, the determinant

$$W = \begin{vmatrix} y_1 & y_2 \\ y_1' & y_2' \end{vmatrix}$$

(16-1)

Wronski showed that, if y_1 and y_2 (on some interval I) are two solutions of the second-order linear homogeneous ODE, then they are linearly independent on I if $W \neq 0$ for at least one point $x \in I$.

Hoëné Wronski
1778-1853
Illustration © 2015, Brandon Ross Sugg

Hour 16. Variation of parameters

Wroński was very ambitious and grandiose. In 1803 he joined the observatory in Marseille France. In 1810 he published a huge work, which he presented as a new foundation of all science and mathematics, after corresponding with almost all of the world's major scientists and mathematicians. It created a lot of controversy and Wronski was subsequently forced out of the observatory. Wroński went on to attempt to build a machine to predict the future and he designed caterpillar-like vehicles that were supposed to replace trains.

This Hour presents another method for finding a particular solution y_p for a non-homogeneous linear ODE. Again, the reason we need to determine a particular solution y_p is so that we can add it on to the homogeneous solution $y_h = C_1 y_1 + C_2 y_2$ in order to form the general solution:

$$y_g = C_1 y_1 + C_2 y_2 + y_p \qquad (16\text{-}2)$$

However, unlike the method of the previous Hour, the present **variation of parameters method** gives a means for finding a particular solution when we have *non-constant coefficients*:

$$y'' + P(x)y' + Q(x)y = g(x) \qquad (16\text{-}3)$$

Note that we have provided for the coefficient in front of y'' to be one (by having divided through by whatever was there when the problem was initially posed). For using variation of parameters, it will be important to start out in this form.

To use the variation of parameters technique, we begin by solving the homogeneous equation $y'' + P(x)y' + Q(x)y = 0$. The homogeneous solution then is:

$$y_h = C_1 y_1 + C_2 y_2 \qquad (16\text{-}4)$$

The basis functions $\{y_1, y_2\}$ of the solution space of the homogeneous equation will be useful in deciding what the particular solution y_p can be.

Next we make the following assumption for the form of y_p:

$$y_p = u_1(x) y_1(x) + u_2(x) y_2(x) \qquad (16\text{-}5)$$

113

Here we have introduced new *functional coefficients* $u_1(x)$ and $u_2(x)$. Eq. 16-5 might look like it "fell out of the sky" but it's actually straightforward to see why it makes sense. We can see that assuming functional coefficients in front of the y's is correct by looking at what is absurd about assuming constant coefficients in front of the y's instead.

If we assume (wrongly) $y_p = A_1 y_1 + A_2 y_2$ with A_1 and A_2 constant, then the left side of $y'' + P(x)y' + Q(x)y = g(x)$ becomes zero. (This is because (A_1, A_2) is a point in the solution space of $y'' + P(x)y' + Q(x)y = 0$.) Stop and think about this for a moment. Thus the constant coefficient assumption forces $g(x)$ to zero which contradicts our initial assumption of having a non-homogeneous equation and $g(x) \neq 0$. Thus we cannot have a particular solution in the form $y_p = A_1 y_1 + A_2 y_2$ with A_1 and A_2 constant; rather, we have to assume functional coefficients $u_1(x)$ and $u_2(x)$ when we express y_p in terms of y_1 and y_2, that is, we must assume $y_p = u_1(x)y_1(x) + u_2(x)y_2(x)$.

The implications of the assumption that y_p can be written as the sum of constants times the y's are as follows:

$$y_p = A_1 y_1 + A_2 y_2 \quad \overset{IMPLIES}{\Rightarrow}$$

$$\frac{d^2}{dx^2}(A_1 y_1 + A_2 y_2) + P(x)\frac{d}{dx}(A_1 y_1 + A_2 y_2) + Q(x)(A_1 y_1 + A_2 y_2) = g(x) \quad \overset{IMPLIES}{\Rightarrow}$$

$$\frac{d^2}{dx^2} A_1 y_1 + P(x)\frac{d}{dx} A_1 y_1 + Q(x) A_1 y_1 +$$
$$\frac{d^2}{dx^2} A_2 y_2 + P(x)\frac{d}{dx} A_2 y_2 + Q(x) A_2 y_2 = g(x) \quad \overset{IMPLIES}{\Rightarrow}$$

Hour 16. Variation of parameters

$$A_1\left(\frac{d^2y_1}{dx^2} + P(x)\frac{dy_1}{dx} + Q(x)y_1\right) +$$

$$A_2\left(\frac{d^2y_2}{dx^2} + P(x)\frac{dy_2}{dx} + Q(x)y_2\right) = g(x) \overset{IMPLIES}{\Rightarrow}$$

$$A_1(0) + A_2(0) = g(x) \overset{IMPLIES}{\Rightarrow} \quad g(x) = 0 \quad (16\text{-}6)$$

This is like saying "a non-homogeneous equation always has the homogeneous term equal to zero" …an idea that is worthy of our contempt! Instead the form we should assume is:

$$y_p = u_1(x)y_1(x) + u_2(x)y_2(x) \quad (16\text{-}7)$$

The above equation is the heart of variation of parameters, a method introduced by Joseph-Louis Lagrange. If we plug Eq. 16-7 into $y'' + Py' + Qy = g(x)$, we get:

$$u_1''y_1 + 2u_1'y_1' + u_1y_1'' + u_2''y_2 + 2u_2'y_2' + u_2y_2'' +$$
$$Pu_1'y_1 + Pu_1y_1' + Pu_2'y_2 + Pu_2y_2' + Qu_1y_1 + Qu_2y_2 = g(x)$$
$$(16\text{-}8)$$

Or:

$$u_1\{y_1'' + Py_1' + Qy_1\} + u_2\{y_2'' + Py_2' + Qy_2\} + P\{u_1'y_1 + u_2'y_2\} +$$
$$\frac{d}{dx}\{u_1'y_1 + u_2'y_2\} + u_1'y_1' + u_2'y_2' = g(x)$$

$$(16\text{-}9)$$

Since y_1 and y_2 are homogeneous solutions, the first two terms reduce to zero:

$$u_1\{0\} + u_2\{0\} + P\{u_1'y_1 + u_2'y_2\} + \frac{d}{dx}\{u_1'y_1 + u_2'y_2\} + u_1'y_1' + u_2'y_2' = g(x)$$

$$(16\text{-}10)$$

Differential Equations in 24 Hours

Let's look carefully at what remains:

$$P\{u_1'y_1 + u_2'y_2\} + (^d/_{dx})\{u_1'y_1 + u_2'y_2\} + u_1'y_1' + u_2'y_2' = g(x)$$

(16-11)

If we let the term in bold equal zero then we have two equations in two unknowns for solving for the u's:

$$u_1'y_1 + u_2'y_2 = 0$$
$$u_1'y_1' + u_2'y_2' = g(x)$$

(16-12)

We are justified in setting $u_1'y_1 + u_2'y_2 = 0$ for the reason that Eq. 16-5 left us enough "wiggle room" in order to do so. Eq. 16-5 represents one equation in the two unknowns u_1 and u_2. But we need to have two equations in the two unknowns in order to determine u_1 and u_2 and thus y_p. Introducing $u_1'y_1 + u_2'y_2 = 0$ gives us the needed second equation, consistent with y_p being a solution to the non-homogeneous equation.

Using Eq. 16-12, we can now solve for the derivatives of u_1 and u_2. We may use the first equation to solve for u_2':

$$u_2' = -u_1'\frac{y_1}{y_2}$$

(16-13)

Plugging this into the second equation, we get:

$$u_1' = -\frac{y_2 g}{(y_1 y_2' - y_1' y_2)}$$

(16-14)

But the denominator is just the Wronskian; therefore,

$$u_1' = -\frac{y_2 g}{W}$$

(16-15)

Hour 16. Variation of parameters

Plugging Eq. 16-15 into Eq. 16-13, we can solve for u_2':

$$u_2' = \frac{y_1 g}{W} \qquad (16\text{-}16)$$

So an algorithm for using variation of parameters is as follows:

Variation of parameters algorithm:

1.) Provide for the differential equation to be in the form $y'' + P(x)y' + Q(x)y = g(x)$.

2.) Find y_1 and y_2, the solutions to the homogeneous equation $y'' + P(x)y' + Q(x)y = 0$.

3.) Calculate W and u_1' and u_2'.

4.) Integrate u_1' and u_2' to form u_1 and u_2. The particular solution is $y_p = u_1 y_1 + u_2 y_2$.

•

With particular solution in hand, the general solution then is just:

$$y_g = y_h + y_p = C_1 y_1 + C_2 y_2 + u_1 y_1 + u_2 y_2$$

$$(16\text{-}17)$$

Example 16-1: Solve:

$$y'' - 2y' + y = 2x \qquad (16\text{-}18)$$

The equation is already in standard form. So the next step is to solve the homogeneous equation:

$$m^2 e^{mx} - 2m e^{mx} + e^{mx} = 0 \qquad (16\text{-}19)$$

This reduces to the characteristic equation $(m-1)(m-1) = 0$ which gives the double root $m = 1$. Thus we have $y_1 = e^x$. For a second solution to the homogeneous equation, let $y_2 = v(x)y_1 = v(x)e^x$. This gives us a second homogeneous solution $y_2 = xe^x$. So the Wronskian is:

Differential Equations in 24 Hours

$$W = \begin{vmatrix} e^x & xe^x \\ e^x & e^x(x+1) \end{vmatrix} = e^{2x}(x+1) - xe^{2x} = e^{2x}$$

(16-20)

We can solve for the derivatives of the u's:

$$u_1' = -\frac{y_2 g}{W} = -\frac{xe^x 2x}{e^{2x}} = -2x^2 e^{-x}$$

$$u_2' = \frac{y_1 g}{W} = \frac{e^x 2x}{e^{2x}} = 2xe^{-x}$$

(16-21)

Now we can integrate and solve for u_1:

$$u_1 = -2\int x^2 e^{-x}\, dx = 2x^2 e^{-x} + 4xe^{-x} + 4e^{-x}$$

(16-22)

We can get u_2 just as easily:

$$u_2 = 2\int xe^{-x}\, dx = 2\left[(-1)\left(xe^{-x} - \int e^{-x}\, dx\right)\right] = -2xe^{-x} - 2e^{-x}$$

(16-23)

Now $y_p = u_1 y_1 + u_2 y_2$ and plugging in we get:

$$y_p = 2x + 4$$

(16-24)

So the general solution is:

$$y = C_1 e^x + C_2 xe^x + 2x + 4$$

(16-25)

◊

Hour 16. Variation of parameters

Joseph-Louis Lagrange
1736 – 1813
Illustration © 2015, Brandon Ross Sugg

Lagrange and Euler were the greatest mathematicians of the 18th century. As mentioned, variation of parameters was developed by Lagrange. Lagrangian mechanics is an incredibly beautiful field which greatly simplifies some complex systems. Lagrange multipliers are of fundamental importance in the calculus of variations. Lagrange also made major contributions in abstract algebra and number theory.

When Lagrange was 56, he married a teenager. As inadvisable as that may be, they apparently had an excellent marriage. We are told that she chased after him and he reluctantly gave in and married her. They were totally devoted to each other and Lagrange couldn't even stand to be away from her while she went shopping.

EXERCISES

Exercise 16-1: Find a particular solution to:

$$y'' + 9y = \tan 3x \qquad (16\text{-}26)$$

Exercise 16-2: Use variation of parameters to solve:

$$y'' - y = e^{2x} \qquad (16\text{-}27)$$

Exercise 16-3: Find the general solution of:

$$y'' + 2y' + y = e^{3x} \qquad (16\text{-}28)$$

Exercise 16-4: Find the general solution of:

$$y'' + y = \csc(x) \qquad (16\text{-}29)$$

Hour 17. Cauchy-Euler Equation

Notice that, in the equation below, the power to which x is raised in each term is the same as the order of derivative of y in that term:

$$a_n x^n \frac{d^n y}{dx^n} + a_{n-1} x^{n-1} \frac{d^{n-1} y}{dx^{n-1}} + \ldots + a_1 x \frac{dy}{dx} + a_0 y = f(x)$$

(17-1)

An equation in the form of Eq. 17-1 is known as a Cauchy-Euler equation. An approach for finding homogeneous solutions (the $f(x) = 0$ case) is to look for solutions in the form:

$$y = x^m \tag{17-2}$$

Let's consider the 2nd-order homogeneous Cauchy-Euler equation:

$$ax^2 \frac{d^2 y}{dx^2} + bx \frac{dy}{dx} + cy = 0 \tag{17-3}$$

Plugging in $y = x^m$, we have:

$$ax^2 m(m-1)x^{m-2} + bxmx^{m-1} + cx^m = 0 \tag{17-4}$$

This becomes:

$$am(m-1)x^m + bmx^m + cx^m = 0 \tag{17-5}$$

Dividing out x^m leaves an auxiliary equation:

$$am^2 + (b-a)m + c = 0 \tag{17-6}$$

How we make use of the m's will give you a sense of déjà vu. For the Cauchy-Euler equation, there are three possible forms the homogeneous solution y_h can take depending on whether the roots of Eq. 17-6 are 1.) distinct real m_1 and m_2, 2.) repeated real m, or 3.) complex conjugate roots $m = \alpha \pm i\beta$.

Differential Equations in 24 Hours

Table 17-1: Homogeneous solution table for 2nd-order Cauchy-Euler equation.

Case 1	Distinct real roots	$y_h = C_1 x^{m_1} + C_2 x^{m_2}$
Case 2	Repeated real root	$y_h = C_1 x^m + C_2 x^m \ln x$
Case 3	Complex roots	$y_h = x^\alpha [C_1 \cos(\beta \ln x) + C_2 \sin(\beta \ln x)]$

Example 17-1: Solve $x^2 y'' + xy' - 9y = 0$.

Solution: Substituting $y = x^m$ we have:

$$x^2 m(m-1)x^{m-2} + xmx^{m-1} - 9x^m = 0 \quad (17\text{-}7)$$

Or:

$$m(m-1)x^m + mx^m - 9x^m = 0 \quad (17\text{-}8)$$

Dividing out x^m provides us the auxiliary equation:

$$m(m-1) + m - 9 = 0 \quad \Rightarrow \quad m^2 - 9 = 0$$

$$(17\text{-}9)$$

Thus we have two distinct real roots, $m_1 = 3$, $m_2 = -3$. So, referring to table 17-1, we see that the solution is:

$$y = C_1 x^3 + C_2 x^{-3} \quad (17\text{-}10)$$

◊

But where does Table 17-1 come from? We can derive Table 17-1 by making the substitution $z = \ln(x)$ in Table 13-1. (Recall that Hour 13 was concerned with finding solutions to equations with constant coefficients.) First replace x

Hour 17. Cauchy-Euler Equation

with z in Table 13-1.

Table: Constant coefficients homogeneous solutions.

Case 1	Distinct real roots	$y_h = C_1 e^{m_1 z} + C_2 e^{m_2 z}$
Case 2	Repeated real root	$y_h = C_1 e^{mz} + C_2 x e^{mz}$
Case 3	Complex roots	$y_h = e^{\alpha z}[C_1 \cos(\beta z) + C_2 \sin(\beta z)]$

Next introduce $z = ln(x)$:

Case 1	Distinct real roots	$y_h = C_1 e^{m_1 \ln x} + C_2 e^{m_2 \ln x}$
Case 2	Repeated real root	$y_h = C_1 e^{m \ln x} + C_2 x e^{m \ln x}$
Case 3	Complex roots	$y_h = e^{\alpha \ln x}[C_1 \cos(\beta \ln x) + C_2 \sin(\beta \ln x)]$

This reduces to Table 17-1 by using some properties of logarithms:

$$e^{m \ln x} = e^{\ln x^m} = x^m$$
$$e^{\alpha \ln x} = e^{\ln x^\alpha} = x^\alpha \tag{17-11}$$

How is it that we can derive the solution table for solutions to the Cauchy-Euler equation so easily from the solution table for the constant coefficients equation? The answer is that $x = e^z$ turns the Cauchy-Euler equation into a constant coefficients equation:

Differential Equations in 24 Hours

$$a\frac{d^2y}{dz^2} + (b-a)\frac{dy}{dz} + cy = 0 \qquad (17\text{-}12)$$

The Cauchy-Euler equation is "a change of variables away" from being a constant coefficient equation, where the change of variables is $x = e^z$ or $z = \ln(x)$. Remember the change of variables is EASY. (Get it? e to the z.)

Example 17-2: Show that the Cauchy-Euler equation $ax^2y'' + bxy' + cy = 0$ reduces to a 2nd-order linear ODE with constant coefficients by making the substitution $x = e^z$.

Solution: Making the substitution $x = e^z$, the derivatives with respect to x can be converted as follows:

$$\frac{dy}{dx} = \frac{dy}{dz}\frac{dz}{dx} = \frac{dy}{dz}e^{-z} \Rightarrow \frac{d^2y}{dx^2} = e^{-2z}\frac{d^2y}{dz^2} - e^{-2z}\frac{dy}{dz}$$

$$(17\text{-}13)$$

Plugging $x = e^z$ and these derivatives into $ax^2y'' + bxy' + cy = 0$, we get:

$$ae^{2z}\left(e^{-2z}\frac{d^2y}{dz^2} - e^{-2z}\frac{dy}{dz}\right) + be^z\left(\frac{dy}{dz}e^{-z}\right) + cy = 0$$

$$(17\text{-}14)$$

Which reduces to:

$$a\frac{d^2y}{dz^2} + (b-a)\frac{dy}{dz} + cy = 0 \qquad (17\text{-}15)$$

Which is a constant coefficient equation.

◊

Let's do another example using Table 17-1.

Hour 17. Cauchy-Euler Equation

Example 17-3: Solve:

$$4x^2 y'' + y = 0 \qquad (17\text{-}16)$$

Solution: Substituting $y = x^m$, we have:

$$4x^2 m(m-1)x^{m-2} + x^m = 0 \quad \Rightarrow \quad 4m(m-1)x^m + x^m = 0 \qquad (17\text{-}17)$$

Whose auxiliary equation is:

$$4m(m-1) + 1 = 0 \quad \Rightarrow \quad m^2 - m + \frac{1}{4} = 0 \qquad (17\text{-}18)$$

Which factors as $(m - \frac{1}{2})^2 = 0$. Thus $m = \frac{1}{2}$ is a double root; therefore, referring to Table 17-1, the solution is:

$$y = C_1 \sqrt{x} + C_2 \sqrt{x} \, (\ln x) \qquad (17\text{-}19)$$

◊

Differential Equations in 24 Hours

Augustin Cauchy
(1789-1857)
Illustration © 2015, Brandon Ross Sugg

Cauchy was born a few years after Euler died, a factor that must have made two-way collaboration on the Cauchy-Euler equation challenging. ☺ Cauchy invented the determinant and contributed enormously to complex analysis, group theory, and many other areas of mathematics.

When Cauchy was a young boy, his family was neighbors with the great mathematician Laplace who dropped in to see the family now and then and encouraged young Cauchy to cultivate his incredible mathematical talent. We shall get to know Laplace later in this book.

When Cauchy was 10, his father was elected Secretary of the Senate. Cauchy's father was given an office in Luxembourg Palace and young Cauchy used a corner of the office as his study. The great mathematician Lagrange stopped by the office frequently to do business with the father but became interested in the

Hour 17. Cauchy-Euler Equation

boy and his amazing mathematical talent. On one occasion Lagrange said, "You see that young man? Well! He will supplant all of us insofar as we are mathematicians."

As a young man, when Cauchy headed off to join the military, he only could bring four books. He brought a book by Laplace, a book by Lagrange, a book about Christ, and a copy of Virgil's works. Despite his military duties, Cauchy carried on mathematical research. In 1811, when he was twenty, he published a major paper on the theory of polyhedra, inaugurating an incredibly prolific mathematical career.

In 1814, Cauchy published a paper on definite integrals with complex-number limits, a major milestone in the development of complex analysis. Soon Cauchy was a member of the Academy of Sciences and then many honorable and important positions came to him quickly. His productivity was incredible. He would often publish two full-length papers in a single week. Some of his papers were as long as 300 pages. The Academy was so swamped by his input that they passed a rule prohibiting the publication of an article longer than four pages.

In his life, Cauchy wrote 789 papers (many of them very massive) filling twenty four large quarto volumes.

Let's solve a Cauchy-Euler equation of the third case, one having an auxiliary equation with complex conjugate roots.

Example 17-4: Solve $x^2 y'' + xy' + 4y = 0$.

Solution: Plugging in $y = x^m$:

$$x^2 m(m-1)x^{m-2} + xmx^{m-1} + 4x^m = 0$$

$$\Downarrow$$

$$m(m-1)x^m + mx^m + 4x^m = 0 \qquad (17\text{-}20)$$

Dividing out x^m and simplifying, we have the auxiliary equation:

$$m^2 + 4 = 0 \qquad (17\text{-}21)$$

This has complex conjugate roots $m = \pm i$. Referring back to Table 17-1, we see

Differential Equations in 24 Hours

that the solution is:

$$y = x^0[C_1 \cos(2 \ln x) + C_2 \sin(2 \ln x)] \qquad (17\text{-}22)$$

Or:

$$y = C_1 \cos(2 \ln x) + C_2 \sin(2 \ln x) \qquad (17\text{-}23)$$

◊

The examples above are for the homogeneous case $f(x) = 0$. For problems where $f(x) \neq 0$ we would proceed as before: First find the homogeneous solution $y_h = C_1 y_1 + C_2 y_2$ and then find a particular solution y_p. The general solution then is the sum of these: $y_g = y_h + y_p$.

EXERCISES

Exercise 17-1: Find the general (homogeneous) solution to $x^2 y'' + 13xy' + 20y = 0$.

Exercise 17-2: Find the general solution to the non-homogeneous equation:

$$x^2 y'' + 3xy' + y = \pi \qquad (17\text{-}24)$$

Exercise 17-3: Find the general solution of $4x^2 y'' + 8xy' + 5y = 0$.

Hour 18. Series Solution

The method of solving differential equations by series solution is about as "brute force" as you can get. The algorithm goes as follows: You simply write down a series,

$$y = \sum_{n=0}^{\infty} c_n x^n \qquad (18\text{-}1)$$

Where the c's are constants and x is the independent variable, and then plug the series into the differential equation at hand. After reducing the result, you determine what formula for generating the c's becomes apparent. Then you also need to establish what interval of convergence--if any--this series has.

The series solution approach can become very tedious to do by hand. The benefit though is that it can be used against very general types of cases such as non-constant coefficient cases. Let's turn to a very easy example to begin to get a feel for how the method works. In the analysis that follows, we will need to keep in mind that the derivative d/dx is a linear operator:

$$\frac{d}{dx}(a_1 f_1(x) + a_2 f_2(x)) = a_1 \frac{d}{dx}(f_1(x)) + a_2 \frac{d}{dx}(f_2(x))$$

$$(18\text{-}2)$$

This will allow us to bring the derivative into a series by allowing it to "pass through" the summation:

$$\frac{d}{dx}\left(\sum_{n=0}^{\infty} c_n x^n\right) = \sum_{n=0}^{\infty} c_n \frac{d}{dx}(x^n) \qquad (18\text{-}3)$$

Example 18-1: Solve $xy + y' = 0$.

Plugging in the series we get:

$$(x)\left(\sum_{n=0}^{\infty} c_n x^n\right) + \frac{d}{dx}\left(\sum_{n=0}^{\infty} c_n x^n\right) = 0 \qquad (18\text{-}4)$$

For the leftmost term, we can use the distributive law to bring the multiplication

Differential Equations in 24 Hours

by x into the series. For the second term, the derivative can be brought into the summation since it is a linear operator.

$$\sum_{n=0}^{\infty} c_n x^{n+1} + \sum_{n=0}^{\infty} c_n \frac{d}{dx}(x^n) = 0 \qquad (18\text{-}5)$$

Which equals:

$$\sum_{n=0}^{\infty} c_n x^{n+1} + \sum_{n=0}^{\infty} c_n n x^{n-1} = 0 \qquad (18\text{-}6)$$

If we can add the two terms appearing on the left, we can arrive at a formula that generates the coefficients. The problem is that the powers of x don't match. Let us write out a few terms of Eq. 18-6 to see what's going on:

$$(c_0 x + c_1 x^2 + ...) + (c_0(0) + c_1 + 2c_2 x^1 + 3c_3 x^2 + ...) = 0$$

$$(18\text{-}7)$$

But this is the same as:

$$(c_0 x + c_1 x^2 + ...) + (c_1 + 2c_2 x + 3c_3 x^2 + ...) = 0$$

$$(18\text{-}8)$$

The subscript on the constants on the left lag by two steps those on the right, from the point of view of desiring to have the powers of x correspond. We can rewrite the second series in Eq. 18-6 so that the powers of x correspond in a way that is conducive to combining the series by "advancing subscripts" in the second series. Note carefully how we rewrite the above--especially the second term. Looking at the second term in Eq. 18-8, we see that replacing n with n+2 and adding the c_1 term "out front" creates exactly the same series as Eq. 18-6 except that now the powers of x are compatible:

$$\sum_{n=0}^{\infty} c_n x^{n+1} + \left(c_1 + \sum_{n=0}^{\infty} (n+2) c_{n+2} x^{n+1} \right) = 0 \qquad (18\text{-}9)$$

Hour 18. Series Solution

We are justified to "bring out" c_1 from the right entry since there is no x-to-the-zero-power term on the left side to match it with. Now we can add the series:

$$\sum_{n=0}^{\infty}(c_n + (n+2)c_{n+2})x^{n+1} + c_1 = 0 \qquad (18\text{-}10)$$

Now, zero, on the right-hand side, is a polynomial (with zero coefficients). Equal polynomials are equal term by term; therefore,

$$c_1 = 0 \quad and \quad c_n + (n+2)c_{n+2} = 0 \;\Rightarrow\; (n+2)c_{n+2} = -c_n$$

$$(18\text{-}11)$$

So we have a **recursion formula** which generates the coefficients:

$$c_{n+2} = -\frac{c_n}{n+2} \qquad (18\text{-}12)$$

Now all of the odd coefficients have to be zero; because if c_1 is zero then c_3 is zero, etc. The even coefficients are a different story. If we start with c_0 we have:

$$c_2 = -\frac{c_0}{2}$$

$$c_4 = -\frac{c_2}{4} = -\frac{1}{4}\left(-\frac{c_0}{2}\right) = \frac{1}{4\times 2}c_0 \qquad (18\text{-}13)$$

$$c_6 = -\frac{1}{6\times 4\times 2}c_0$$

Etc.

We recognize a pattern:

$$c_{2n} = (-1)^n \frac{2^{-n}}{n!} c_0 \qquad (18\text{-}14)$$

Differential Equations in 24 Hours

Thus the solution is:

$$y = \sum_{n=0}^{\infty} c_n x^n \qquad (18\text{-}15)$$

Or:

$$y = c_0 \sum_{n=0}^{\infty} (-1)^n \frac{2^{-n}}{n!} x^{2n} \qquad (18\text{-}16)$$

Or:

$$y = c_0 \sum_{n=0}^{\infty} \frac{1}{n!}\left(-\frac{x^2}{2}\right)^n \qquad (18\text{-}17)$$

But the Maclaurin expansion for the exponential is:

$$\sum_{n=0}^{\infty} \frac{1}{n!} \varphi^n = e^\varphi \qquad (18\text{-}18)$$

So $\varphi = -\tfrac{1}{2} x^2$ and the final answer is:

$$y = c_0 e^{-x^2/2} \qquad (18\text{-}19)$$

Which converges for all x.

◊

This example suggests an algorithm for applying series solution:

Algorithm for Series Solution:

1.) Plug in the series $y = \sum_{n=0}^{\infty} c_n x^n$.

2.) Distribute coefficients and derivatives so that they appear within the summations.

Hour 18. Series Solution

3.) Advance indices as needed so that powers of x correspond if multiple summations appear. (It may be helpful to write out a few terms of the summations to understand how to perform this. And this step may result in constant terms appearing as separate summands which do not participate in any of the "sigma summations.")

4.) Combine multiple summations by adding like terms.

5.) Determine coefficient generating function(s) by matching like terms on the left and right sides of the equation.

6.) Try to reduce the coefficient generating functions to recursion relations.

7.) Try to recognize any closed form functions that the recursion relations may correspond to. •

Let's apply our algorithm to the solution of an equation which is extremely important to engineering applications. Bessel's equation of order p, where p is a non-negative constant is:

$$x^2 y'' + x y' + (x^2 - p^2) y = 0 \qquad (18\text{-}20)$$

Let's restrict our attention to the zeroeth-order case:

$$x^2 y'' + x y' + (x^2 - p^2) y = 0 \qquad (18\text{-}21)$$

$$x^2 y'' + x y' + x^2 y = 0 \qquad (18\text{-}22)$$

Plugging in the series substitution we obtain:

$$x^2 \sum_{n=0}^{\infty} n(n-1) c_n x^{n-2} + x \sum_{n=0}^{\infty} n c_n x^{n-1} + x^2 \sum_{n=0}^{\infty} c_n x^n = 0$$

$$(18\text{-}23)$$

Bringing the coefficient functions into the summations, we have:

Differential Equations in 24 Hours

$$\sum_{n=0}^{\infty} n(n-1)c_n x^n + \sum_{n=0}^{\infty} nc_n x^n + \sum_{n=0}^{\infty} c_n x^{n+2} = 0$$

(18-24)

Writing out a few terms, we see that:

$$(2c_2 x^2 + 3 \times 2c_3 x^3 + ...) + (c_1 x + 2c_2 x^2 + ...) + (c_0 x^2 + c_1 x^3 + ...) = 0$$

(18-25)

If we advance exponents in the left two summands in Eq. 18-24, we have, being careful not to discard the $c_1 x$ term:

$$\sum_{n=0}^{\infty}(n+2)(n+1)c_{n+2} x^{n+2} + c_1 x +$$

$$\sum_{n=0}^{\infty}(n+2)c_{n+2} x^{n+2} + \sum_{n=0}^{\infty} c_n x^{n+2} = 0$$

(18-26)

These series can now be added:

$$\sum_{n=0}^{\infty} \{(n+2)(n+1)c_{n+2} + (n+2)c_{n+2} + c_n\} x^{n+2} + c_1 x = 0$$

(18-27)

Or:

$$\sum_{n=0}^{\infty} \{(n+2)^2 c_{n+2} + c_n\} x^{n+2} + c_1 x = 0$$

(18-28)

On the right-hand side of Eq. 18-28 we have the polynomial with all coefficients zero. Thus, comparing like terms on the left and right of Eq. 18-28, we have:

$$c_1 = 0$$
$$(n+2)^2 c_{n+2} + c_n = 0 \qquad (18\text{-}29)$$

This gives the coefficient generator:

$$c_1 = 0$$
$$c_0 = c_0$$
$$c_{n+2} = \frac{-c_n}{(n+2)^2} \qquad (18\text{-}30)$$

Which generates coefficients for the even-powered terms only. (All of the odd-powered terms are zero since $c_1 = 0$.) If we use this generating function to produce a few terms, we begin to recognize a pattern:

$$c_0 = c_0 = \frac{c_0}{(0!)^2 (2^2)^0}$$

$$c_2 = \frac{-c_0}{2^2} = (-1)\frac{c_0}{(2 \times 1)^2} = (-1)\frac{c_0}{(1!)^2 (2^2)^1}$$

$$c_4 = \frac{-c_2}{4^2} = \frac{c_0}{4^2 2^2} = \frac{c_0}{(2 \times 2)^2 (2 \times 1)^2} = \frac{c_0}{(2!)^2 (2^2)^2}$$

$$c_6 = \frac{-c_4}{6^2} = \frac{-c_0}{6^2 4^2 2^2} = (-1)\frac{c_0}{(3 \times 2)^2 (2 \times 2)^2 (2 \times 1)^2} =$$

$$(-1)\frac{c_0}{(3!)^2 (2^2)^3}$$

...

$$(18\text{-}31)$$

$$c_{2n} = (-1)^n \frac{c_0}{(n!)^2 2^{2n}} \qquad (18\text{-}32)$$

So the series solution then is:

$$y = c_0 \sum_{n=0}^{\infty} (-1)^n \frac{1}{(n!)^2 2^{2n}} x^{2n}$$

(18-33)

◊

Functions such as this one that satisfy Bessel's equation are called **Bessel functions**. For applications, much of the interest in differential equations is focused on the functions they produce. Equations like Bessel's equation produce families of functions that can be used to represent engineering functions and signals in a way that greatly reduces the difficulty of engineering problems. For example, problems involving propagation of waves from a source having cylindrical symmetry are often greatly simplified when expressed using Bessel functions.

Hour 18. Series Solution

Daniel Bernoulli
(1700-1782)
Illustration © 2015, Brandon Ross Sugg

Bessel functions were first defined by Daniel Bernoulli. Daniel Bernoulli was a son of the Swiss mathematician John Bernoulli whom we have already mentioned. Daniel was close friends with Euler.

Daniel Bernoulli made major contributions to calculus, differential equations, probability, the theory of vibrating strings, a kinetic theory of gases, and many other problems in applied mathematics. Daniel Bernoulli has been called the founder of mathematical physics.

Once when travelling as a young man Daniel modestly introduced himself to an interesting stranger with whom he had been talking: "I am Daniel Bernoulli" "And I," said the other sarcastically, "am Isaac Newton." Daniel was delighted by this to the end of his life.

Exercise 18-1: Use the Series Solution to find the general solution to $y'' + 49y = 0$.

Exercises 18-2 through 18-4 refer to Hermite's equation:

$$y'' - 2xy' + 2py = 0 \qquad (18\text{-}34)$$

The first three Hermite functions are 1, x, and $1 - 2x^2$.

Exercise 18-2: Find a solution of the p = 0 case of Hermite's equation.

Exercise 18-3: Find a solution of the p = 1 case of Hermite's equation.

Exercise 18-4: Find a solution of the p = 2 case of Hermite's equation.

Charles Hermite
(1822-1901)
Illustration © 2015, Brandon Ross Sugg

Charles Hermite was born in north-eastern France December 24, 1822. Hermite studied at the prestigious École Polytechnique for one year; however, he was not permitted to continue owing to a defect in his right foot. He studied on his own for several years and then took and passed his baccalauréat exam. He was then appointed répétiteur and admissions examiner at École Polytechnique.

In 1856, Hermite caught smallpox. Cauchy provided strong religion to help Hermite through this difficulty, converting him to Roman Catholicism. Hermite made important contributions to number theory, algebra, orthogonal polynomials, and elliptic functions. Picard was a son-in-law of Hermite.

Part three:
Laplace Transforms

Hour 19.   Introduction to the Laplace Transform 143

Hour 20.   Introduction to the Laplace Transform Exercises 163

Hour 21.   Translation Theorems of Laplace Transforms 165

Hour 22.   Translation Theorems of Laplace Transforms Exercises 183

Hour 19. Introduction to the Laplace Transform

For the next two sections, we will be discussing **transform methods** for solving problems involving differential equations. Up to this point, the kinds of methods we have been using have been *in situ* techniques. "In situ" is Latin meaning "in place." For the most part, the problems we have seen so far have been solved *in place* by plugging in substitutions which result in simplifications that, in turn, eventually result in solution. For the transform methods that we take up presently, our philosophy will be quite different: Rather than solve the problem *in situ*, we will *transform* the problem into a new domain--a realm in which the problem is more tractable--and work on it in that new domain. Then once the problem is solved in the transform domain, we will transform the answer back to the original realm.

Once we solve the problems in the transform domain, we will **inverse transform** (also called reverse transform) the solutions back to the original problem domain and report the answer.

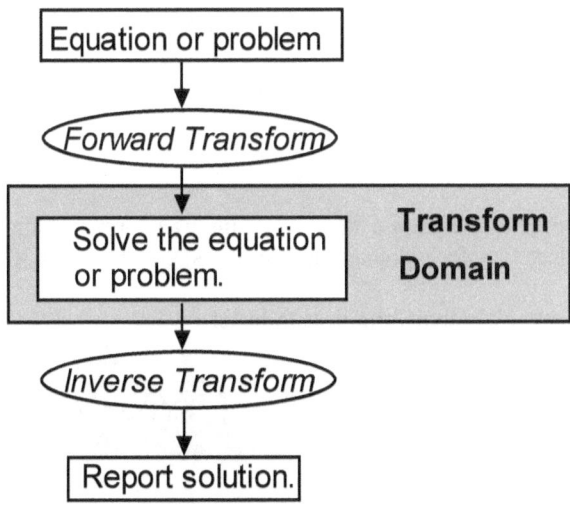

Figure 19-1: Transform Approach to Problem Solving.

The **Laplace transform** will be the particular type of transform that we consider. We will take problems involving differential equations (that would be

Differential Equations in 24 Hours

hard to solve *in situ*) and transform them into the Laplace domain and find that they are easy to tackle in the Laplace domain. Once we have solved the problems in the Laplace domain, we will inverse-Laplace transform these problems back to the original setting of the problem.

We turn now to the definition of a Laplace transform:

Definition 19-1: The **Laplace transform** of a function f(t) is the inner product between f(t) and an exponential function of the form e^{-st}:

$$\mathcal{L}(f(t)) = \int_0^\infty f(t)e^{-st}dt \tag{19-1}$$

■

Note that the integration is from zero to infinity. The result of this integration is to produce a function of s, that is; t is only a dummy variable; therefore, t is only used to accomplish the integration and then is discarded. The original function is a function of t, in particular *f(t)*; however, in the transform domain we have a function of s. This rendering of *f* in the transform domain is written *F(s)*.

$$F(s) = \int_0^\infty f(t)e^{-st}dt \tag{19-2}$$

The capitol letter indicates a function in the transform domain. Therefore, y(t) in the Laplace transform domain becomes Y(s), x(t) becomes X(s), and so on.

The inverse Laplace transform is as follows:

$$f(t) = \frac{1}{2\pi i}\int_{c-i\infty}^{c+i\infty} F(s)e^{ts}ds \tag{19-3}$$

Which is true for c "large enough." It is rare for this latter integral, discovered by Poisson, to be used directly.

A function in the input domain and its counter-part in the transform domain are referred to as transform pairs. The inverse transform is usually never calculated because Eq. 19-2 is sufficient to generate the needed *transform pairs*. That is, all of the means for connecting a function to its Laplace domain counterpart is provided by Eq. 19-2.

Hour 19. Introduction to the Laplace Transform

Let's now take a moment to see what some familiar functions become when they are transformed into the Laplace domain. First, let's determine the Laplace transform of unity.

$$\mathcal{L}(1) = \int_0^\infty 1\, e^{-st} dt \Rightarrow$$

$$-\frac{1}{s}\int_0^\infty e^{-st}(-s)dt = -\frac{1}{s} e^{-st} \Big|_0^\infty = -\frac{1}{s}(0-1) = \frac{1}{s}$$

(19-4)

So y(t) = 1 in the transform domain becomes Y(s) = 1/s in the Laplace domain.

The Laplace transform of t is also not difficult.

$$\mathcal{L}(t) = \int_0^\infty t\, e^{-st} dt \Rightarrow -\frac{1}{s}\int_0^\infty t\, e^{-st}(-s)dt$$

(19-5)

We can use integration by parts as follows:

Let $u = t$ and $dv = e^{-st}(-s)dt$

$$\mathcal{L}(t) = -\frac{1}{s} e^{-st} t \Big|_0^\infty - \left(-\frac{1}{s}\int_0^\infty e^{-st} dt\right)$$

$$= 0 + \frac{1}{s}\int_0^\infty e^{-st} dt = \frac{1}{s}\mathcal{L}(1) = \frac{1}{s^2}$$

(19-6)

Thus y(t) = t in the "input domain" becomes Y(s) = 1/s² in the Laplace domain.

You might have noticed that we used the result for the transform of *1* to calculate the transform of *t*. In general the Laplace transform of t^n has a very

Differential Equations in 24 Hours

simple relationship to the Laplace transform of t^{n-1}.

$$\mathcal{L}(t^n) = \frac{n}{s}\mathcal{L}(t^{n-1}) \tag{19-7}$$

This is easy to derive using integration by parts:

$$u = t^n \qquad dv = -\frac{1}{s}e^{-st}(-s)dt \qquad \text{and}$$

$$du = nt^{n-1}dt \qquad v = -\frac{1}{s}e^{-st}$$

$$\mathcal{L}(t^n) = \int_0^\infty t^n e^{-st} dt = -t^n \frac{1}{s}e^{-st}\Big|_0^\infty - \int_0^\infty \left(-\frac{1}{s}\right)e^{-st} nt^{n-1} dt$$

$$\tag{19-8}$$

The exponential e^{st}, for large t, increases faster than any polynomial, including t^n. Thus $\lim_{t \to \infty} -\frac{t^n}{e^{st}}\frac{1}{s} = 0$. The first term on the right of the equals sign is then zero. Thus:

$$\mathcal{L}(t^n) = 0 + \frac{n}{s}\int_0^\infty t^{n-1} e^{-st} dt = \frac{n}{s}\mathcal{L}(t^{n-1}) \tag{19-9}$$

Using Eq. 19-6, we can generate the Laplace transforms of all of the powers of t:

$$\mathcal{L}(t^2) = \frac{2}{s}\mathcal{L}(t) = \frac{2}{s}\frac{1}{s} = \frac{2}{s^2}$$

$$\mathcal{L}(t^3) = \frac{3}{s}\mathcal{L}(t^2) = \frac{3}{s}\left(\frac{2}{s^2}\right) = \frac{3!}{s^3}$$

$$\vdots$$

$$\mathcal{L}(t^n) = \frac{n!}{s^{n+1}} \tag{19-10}$$

146

Hour 19. Introduction to the Laplace Transform

The Laplace transform of e^{at} is easy to produce in a two-line derivation:

$$L(e^{at}) = \int_0^\infty e^{at} e^{-st} dt = \int_0^\infty e^{(a-s)t} dt =$$

$$\frac{1}{a-s} \{ e^{(a-s)t} \big|_0^\infty = \frac{1}{s-a}$$

(19-11)

The Laplace transform of sin(at) can be arrived at by integrating by parts twice. The first integration by parts gives us an inner product with cosine:

$$\mathcal{L}(\sin(at)) = \int_0^\infty \sin(at) e^{-st} dt$$

$$u = \sin(at) \qquad dv = e^{-st} dt$$

$$du = a\cos(at) \qquad v = -\frac{1}{s} e^{-st}$$

$$\mathcal{L}(\sin(at)) = -\sin(at)\frac{1}{s} e^{-st} \big|_0^\infty + \frac{a}{s} \int_0^\infty e^{-st} \cos(at) dt$$

$$= \frac{a}{s} \int_0^\infty e^{-st} \cos(at) dt$$

(19-12)

A second integration by parts gives us two terms involving the inner product with sine:

147

Differential Equations in 24 Hours

$$u = \cos(at) \qquad dv = e^{-st}\,dt$$

$$du = -a\sin(at)\,dt \qquad v = -\frac{1}{s}e^{-st}$$

$$L(\sin(at)) = \frac{a}{s}[\,-\frac{1}{s}\cos(at)e^{-st}\,\Big|_0^\infty -$$

$$\int_0^\infty (-\frac{1}{s})e^{-st}(-a)\sin(at)\,dt\,]$$

$$= \frac{a}{s}[\,0 - (-\frac{1}{s}) - \frac{a}{s}\int_0^\infty e^{-st}\sin(at)\,dt\,]$$

$$= \frac{a}{s^2} - \frac{a^2}{s^2}\int_0^\infty e^{-st}\sin(at)\,dt$$

(19-13)

The second term on the right of the equals sign involves $L(\sin(at))$.

$$L(\sin(at)) = \frac{a}{s^2} - \frac{a^2}{s^2}\int_0^\infty e^{-st}\sin(at)\,dt =$$

$$\frac{a}{s^2} - \frac{a^2}{s^2}L(\sin(at))$$

(19-14)

If we combine terms involving $L(\sin(at))$ we arrive at:

$$L(\sin(at)) = \frac{a}{s^2} - \frac{a^2}{s^2}L(\sin(at)) \Rightarrow$$

$$L(\sin(at))\{1 + \frac{a^2}{s^2}\} = \frac{a}{s^2}$$

(19-15)

Thus the Laplace transform of $\sin(at)$ is:

Hour 19. Introduction to the Laplace Transform

$$\mathcal{L}(\sin(at)) = \frac{\frac{a}{s^2}}{\left\{1 + \frac{a^2}{s^2}\right\}} = \frac{a}{s^2 + a^2}$$

(19-16)

When we were performing the above derivation, Eq. 19-12 surreptitiously gave us a relationship between the Laplace transform of cos(at) and the Laplace transform of sin(at).

$$\mathcal{L}(\sin(at)) = \frac{a}{s}\int_0^\infty e^{-st}\cos(at)\,dt = \frac{a}{s}\mathcal{L}(\cos(at))$$

(19-17)

We can use this immediately to obtain $\mathcal{L}(\cos(at))$:

$$\mathcal{L}(\cos(at)) = \frac{s}{a}\mathcal{L}(\sin(at)) = \frac{s}{a}\left(\frac{a}{s^2 + a^2}\right) = \frac{s}{s^2 + a^2}$$

(19-18)

The Laplace transform is a linear transform. This means that the Laplace transform of a sum of functions is the sum of the Laplace transforms of the individual functions. Linearity also means that the Laplace transform of a constant times a function is the constant times the Laplace transform of the function. We can express both aspects of the Linearity of the Laplace transform in a single equation:

$$\mathcal{L}(c_1 y_1(t) + c_2 y_2(t)) = c_1\mathcal{L}(y_1(t)) + c_2\mathcal{L}(y_2(t))$$

(19-19)

We can use this property in the development of the Laplace transforms of sinh(at) and cosh(at). In the Laplace domain, the hyperbolic sine and the hyperbolic cosine differ from their non-hyperbolic counterparts by having a minus sign between s² and a² in the denominator:

$$L(\sinh(at)) = \int_0^\infty \frac{e^{at} - e^{-at}}{2} e^{-st} dt$$

$$= \frac{1}{2}\int_0^\infty e^{at}e^{-st} dt - \frac{1}{2}\int_0^\infty e^{-at}e^{-st} dt$$

$$= \frac{1}{2}L(e^{at}) - \frac{1}{2}L(e^{-at}) \qquad (19\text{-}20)$$

$$= \frac{1}{2}\left\{\frac{1}{s-a} - \frac{1}{s+a}\right\}$$

$$= \frac{a}{s^2 - a^2}$$

$$\mathcal{L}(\cosh(at)) = \int_0^\infty \frac{e^{at} + e^{-at}}{2} e^{-st} dt$$

$$= \frac{1}{2}\mathcal{L}(e^{at}) + \frac{1}{2}\mathcal{L}(e^{-at}) = \frac{1}{2}\left\{\frac{1}{s-a} + \frac{1}{s+a}\right\}$$

$$= \frac{s}{s^2 - a^2}$$

$$(19\text{-}21)$$

We can now tabulate the Laplace transforms we have developed so far and form an initial picture of how some familiar functions appear in the Laplace domain.

Hour 19. Introduction to the Laplace Transform

Table 19-1: Basic Laplace Transforms.

Input Domain	Laplace Domain
0	0
1	$\dfrac{1}{s}$
t	$\dfrac{1}{s^2}$
t^n	$\dfrac{n!}{s^{n+1}}$
e^{at}	$\dfrac{1}{s-a}$
sin(at)	$\dfrac{a}{s^2+a^2}$
cos(at)	$\dfrac{s}{s^2+a^2}$
sinh(at)	$\dfrac{a}{s^2-a^2}$
cosh(at)	$\dfrac{s}{s^2-a^2}$

It's interesting that all of the various types of functions on the left (polynomials, trigonometric functions, and hyperbolic functions) become ratios of polynomials when transformed to the Laplace domain. These varied functions seem to be on more of a common footing in the Laplace domain, being rational functions in the Laplace domain.

Each row of the table above is an example of a transform pair. The left element is a function in the input domain. The right element is how that function manifests in the

Laplace domain, for example, $\left(\cos(at), \dfrac{s}{s^2 + a^2}\right)$ is a transform pair.

Solving problems involving Laplace transforms entails keeping "two sets of books," one to account for the behavior in the input domain and the other to account for the behavior in the Laplace domain. One can think of Table 19-1 as two accounting books laid side by side. Although we could use Eq. 19-2 and Eq. 19-3 to transform functions back and forth between the input and Laplace domains, we will usually not use Eq. 19-2 and we will almost never use Eq. 19-3. Instead, most of the time we can make the transitions between the two domains by looking up the transform pairs in tables such as Table 19-1.

To illustrate the point, let's use Table 19-1 to transform some other familiar functions into the Laplace domain.

Example 19-1: Determine the Laplace transform of a line.

Solution:

$$y = at + b$$

$$\mathcal{L}(at + b) = a\mathcal{L}(t) + b\mathcal{L}(1) = a\dfrac{1}{s^2} + b\dfrac{1}{s}$$

$$Y(s) = a\dfrac{1}{s^2} + b\dfrac{1}{s}$$

(19-22)

◊

Hour 19. Introduction to the Laplace Transform

Here we have transformed a whole equation to the Laplace domain. The equation y = at + b becomes $Y(s) = a\dfrac{1}{s^2} + b\dfrac{1}{s}$ when it is transformed to the Laplace domain.

In the next example, the sum of a trigonometric function and a hyperbolic function becomes a simple rational function in the Laplace domain.

Example 19-2: y = cos(at) + sinh(at)

Solution:
$$L(y) = L(\cos(at)) + L(\sinh(at))$$
$$\Downarrow$$

$$Y(s) = \frac{s}{s^2 + a^2} + \frac{a}{s^2 - a^2}$$
$$= \frac{s(s^2 - a^2)}{s^4 - a^4} + \frac{a(s^2 + a^2)}{s^4 - a^4} = \frac{s^3 + sa(s-a) + a^3}{s^4 - a^4}$$

(19-23)

◊

In many instances, Laplace transforms make easy work out of dealing with some rather difficult functions. Functions containing a discontinuity often present challenges. Some discontinuous functions are easy to represent and work with in the Laplace domain, provided they are piecewise continuous. For instance, we can take the Laplace transform of a piecewise defined step. Before we look at any particular step, we need to introduce the unit step function u(t). The function u(t) is equal to zero for t < 0. It is equal to one for t ≥ 0. This allows you to represent a unit step that starts at t = a by u(t - a) since t = a makes the argument of *u* zero and thus initiates the start of the upper part of the step. Consider the function:

$$f(t) = \begin{cases} 0, & 0 \le t < a \\ b, & a \le t < \infty \end{cases}$$

(19-24)

Differential Equations in 24 Hours

It can be represented as $b\,u(t - a)$ and appears as follows:

The Laplace transform of this function is:

$$\mathcal{L}(f(t)) = \int_0^a 0\, dt + \int_a^\infty be^{-st} dt = \frac{b}{s}e^{-as} \qquad (19\text{-}25)$$

The trick here was to break the original integral, which was from zero to infinity, into an integral from zero to a followed by an integral from a to infinity. The final result is an amazingly harmless-looking representation of a discontinuous function. What's amazing is that all of the information about the step (start time, the step height, and the shape of the step) is contained in the single term $(b/s)e^{-as}$. This is accomplished without any piecewise definitions. The fact that the discontinuous square wave is transformed to a simple exponential function suggests that problems involving such functions may be much easier to solve in the Laplace domain.

Perhaps the most powerful feature of the Laplace transform is the way it transforms derivatives of functions. Differentiation of y(t) in the input domain manifests as multiplication of Y(s) by s in the Laplace domain with the additional chore of subtracting by y(0).

Hour 19. Introduction to the Laplace Transform

Table 19-2: Laplace Transform of a Derivative.

Input Domain	Laplace Domain
$\dfrac{dy}{dt}$	$sY(s) - y(0)$

The term being subtracted off is lower-case y(0), which is the <u>input domain</u> initial condition. The object sY(s) - y(0) lives in the Laplace domain notwithstanding the appearance of y(0) which is lower-case. This presents no conceptual problem since y(0) is only a constant, being evaluated at a fixed value of the independent variable.

The next example illustrates using the Laplace transform to solve an initial value problem. Notice how the derivative is reduced to a multiplication by *s* and a subtraction of *y(0)*. The first step is to Laplace-transform the whole differential equation.

Example 19-3: Solve the initial value problem: y' - y = 2cos(5t) with y(0) = 0.

Solution: Applying the Laplace transform to the differential equation, we have:

$$sY(s) - y(0) - Y(s) = 2\left(\frac{s}{s^2 + 5^2}\right) \tag{19-26}$$

Plugging in the initial condition y(0) = 0, we have:

$$sY(s) - Y(s) = \frac{2s}{s^2 + 5^2}$$

$$Y(s)(s - 1) = \frac{2s}{s^2 + 5^2} \tag{19-27}$$

Differential Equations in 24 Hours

This can be solved for Y(s):

$$Y(s) = \frac{2s}{(s-1)(s^2+5^2)} \tag{19-28}$$

Using partial fractions, we can decompose the right-hand side of this into easier terms:

$$\frac{2s}{(s-1)(s^2+5^2)} = \frac{As}{(s^2+5^2)} + \frac{B}{(s^2+5^2)} + \frac{C}{(s-1)} \tag{19-29}$$

Multiplying through by $(s-1)(s^2+5^2)$, we have:

$$2s = As(s-1) + B(s-1) + C(s^2+5^2)$$
$$2s = (A+C)s^2 + (B-A)s - B + C5^2 \tag{19-30}$$

Comparing like terms, this gives three equations in three unknowns:

$$\begin{array}{rcl} A + C & = & 0 \\ -A + B & = & 2 \\ -B + 25C & = & 0 \end{array} \tag{19-31}$$

Whose solution is: $A = \frac{-1}{13}$, $B = \frac{25}{13}$, and $C = \frac{1}{13}$.

Equation 19-28 then becomes:

$$Y(s) = \left(-\frac{1}{13}\right)\frac{s}{(s^2+5^2)} + \left(\frac{5}{13}\right)\frac{5}{(s^2+5^2)} + \left(\frac{1}{13}\right)\frac{1}{(s-1)} \tag{19-32}$$

Hour 19. Introduction to the Laplace Transform

Referring to table 19-1, the terms on the right-hand side can be matched to their time domain equivalents. Thus we can inverse-Laplace transform Eq. 19-42 to obtain the solution:

$$y(t) = -\frac{1}{13}\cos(5t) + \frac{5}{13}\sin(5t) + \frac{1}{13}e^t \quad (19\text{-}33)$$

◊

To derive the Laplace transform of a derivative formula, we use integration by parts to put the derivative onto the exponential:

$$\mathsf{L}\left(\frac{df}{dt}\right) = \int_0^\infty \frac{df}{dt} e^{-st}\, dt =$$

$$f(t)e^{-st}\Big|_0^\infty - \int_0^\infty f(t)(-s e^{-st})\, dt =$$

$$0 - f(0) - (-s)\int_0^\infty f(t)e^{-st}\, dt = s\mathsf{L}(f) - f(0)$$

$$(19\text{-}34)$$

Again, the primary effect of differentiation is multiplication by s in the Laplace domain. Then the secondary effect is subtraction of the original function evaluated at zero.

The Laplace transform of the second derivative can be derived using the Laplace transform of the first derivative:

$$\mathsf{L}\left(\frac{d^2 f}{dt^2}\right) = s\mathsf{L}\left(\frac{df}{dt}\right) - \frac{df}{dt}(0)$$

$$= s\bigl(s\mathsf{L}(f) - f(0)\bigr) - \frac{df}{dt}(0) \quad (19\text{-}35)$$

$$= s^2\mathsf{L}(f) - s f(0) - \frac{df}{dt}(0)$$

Now we have two derivatives that we can tabulate.

Differential Equations in 24 Hours

Table 19-3: *Laplace Transforms of Derivatives.*

Input Domain	Laplace Domain
$\dfrac{dy}{dt}$	$sY(s) - y(0)$
$\dfrac{d^2y}{dt^2}$	$s^2Y(s) - sy(0) - y'(0)$

Solving the harmonic oscillator problem was quite a chore using the *in situ* methods of the previous Hours. Using the Laplace transform, it's a breeze.

Example 19-4: Harmonic oscillator. Solve $\ddot{y} + \omega^2 y = 0$ with $y(0) = 1$ and $y'(0) = 0$.

(ω is a constant parameter, "frequency.")

Solution: Apply the Laplace transform to the entire equation:

$$\mathcal{L}\left(\frac{d^2y}{dt^2} + \omega^2 y\right) = \mathcal{L}(0) \tag{19-36}$$

The next step is to use the linearity of the Laplace transform to distribute $\mathcal{L}$.

$$\mathcal{L}\left(\frac{d^2y}{dt^2}\right) + \omega^2 \mathcal{L}(y) = \mathcal{L}(0) \tag{19-37}$$

Making use of our tables, we can write the Laplace domain equation:

$$s^2Y(s) - sy(0) - y'(0) + \omega^2 Y(s) = 0 \tag{19-38}$$

Using the initial conditions $y(0) = 1$ and $y'(0) = 0$, we have:

$$s^2Y(s) - s + \omega^2 Y(s) = 0 \tag{19-39}$$

Hour 19. Introduction to the Laplace Transform

We can solve this for Y(s):

$$Y(s) = \frac{s}{s^2 + \omega^2} \tag{19-40}$$

We have solved the problem in the Laplace domain by isolating the independent variable Y and expressing in in terms of the dependent variable s. In other words, we have found Y = Y(s). Now we need to transform our answer back to the "real world" domain where it appears in the form y = y(t). If we look at Table 19-1, we see that $\frac{s}{s^2 + a^2}$ corresponds to cos(at) in the input domain

Thus the inverse-transform of $Y(s) = \frac{s}{s^2 + \omega^2}$ is cos(ωt).

$$y(t) = \cos(\omega t) \tag{19-41}$$

◊

Being able to effect differentiation via multiplication goes both ways. We can take a derivative with respect to s in the Laplace domain by multiplying by -t in the input domain:

$$\mathcal{L}(-t\, f(t)) = \frac{d}{ds} F(s) \tag{19-42}$$

Leibniz's rule allows us to take the derivative into the integral sign when we differentiate the following with respect to s:

$$F(s) = \int_0^\infty f(t) e^{-st} dt \tag{19-43}$$

Thus we have:

$$\frac{d}{ds} F(s) = \int_0^\infty \frac{\partial}{\partial s} f(t) e^{-st} dt = \int_0^\infty f(t)(-t) e^{-st} dt = \mathcal{L}(-t\, f(t)) \tag{19-44}$$

Differential Equations in 24 Hours

We can use this to determine the Laplace transforms of some harder functions.

Example 19-5: Determine the Laplace transform of $-t\cos(at)$.

The Laplace transform of $\cos(at)$ is:

$$\frac{s}{s^2 + a^2} \qquad (19\text{-}45)$$

Multiplication by $-t$ causes differentiation in the Laplace domain:

$$\frac{1}{s^2 + a^2} + s(-1)\frac{2s}{\left(s^2 + a^2\right)^2} \qquad (19\text{-}46)$$

Which becomes:

$$\frac{a^2 - s^2}{\left(s^2 + a^2\right)^2} \qquad (19\text{-}47)$$

◊

Hour 19. Introduction to the Laplace Transform

Pierre-Simon Laplace
1749-1827
Illustration © 2015, Brandon Ross Sugg

Pierre-Simon Laplace had a profound influence on mathematics, astronomy, and physics and wrote huge texts on these subjects. Laplace's great works featured not only Laplace's contributions to math and science, but also results developed by other geniuses. Unfortunately, Laplace sometimes presented other mathematicians' material in a way that suggested the offerings were all his own. Notwithstanding the problem of omitted acknowledgments, the contributions of Laplace to science were enough for him to be remembered as "The Newton of France." Also there was another side of Laplace that was extremely generous. Jean-Baptiste Biot, as a young man, presented a paper at a venue where Laplace was present. Laplace took Biot aside afterwards and showed him an yellowed unpublished manuscript proving that he (Laplace) had discovered the same result much earlier. Laplace told Biot to go ahead and publish the result and take credit for it.

Differential Equations in 24 Hours

As a young man Laplace made himself famous by proving, for the first time, that the solar system was stable: The planets won't wander off or come crashing together to destroy the solar system. This was a staggering accomplishment because even Newton thought that the solar system might need divine intervention from time to time for the planets to maintain their orbits.

Starting with an idea from Lagrange, Laplace developed the mathematics of the potential which underpins so many areas of science including electromagnetism and gravitation. Use of potential functions allows one to replace partial differential equations in two or three unknowns with an equation in a single unknown.

One of the great tools that was developed by Laplace is the Laplace transform, which was first published in 1782. This transform was used extensively in a huge treatise in probability that Laplace wrote.

Hour 20. Introduction to the Laplace Transform Exercises

Exercise 20-1: Determine Y(s), the Laplace transform of the following time domain function y(t):

$$y(t) = 2\sin(t)\cos(t) \tag{20-1}$$

Exercise 20-2: Determine Y(s), the Laplace transform of y(t), where y(t) is as follows:

$$y(t) = t^6 \sin^2(7t) + t^6 \cos^2(7t) \tag{20-2}$$

Exercise 20-3: Determine y(t), the <u>inverse</u> Laplace transform of Y(s):

$$Y(s) = \frac{1}{s} - \frac{1}{s^2 + 1} \tag{20-3}$$

Exercise 20-4: Use Laplace transforms to solve the following initial value problem.

$$2\frac{dy}{dt} + y = 0 \qquad y(0) = -3. \tag{20-4}$$

Exercise 20-5: Solve the initial value problem:

$$y'' + 2y' + 2y = 2 \quad \text{with} \quad y(0) = 1 \text{ and } y'(0) = 0 \tag{20-5}$$

Exercise 20-6: Solve $\ddot{y} + \omega^2 y = 0$ with y(0) = 0 and y'(0) = 1.

Exercise 20-7: A step function is defined as in Equation 19-24:

Determine the Laplace transform of the *derivative* of this function. Let b = 1.

Exercise 20-8: Determine the Laplace transform of $t^2 sin(at)$.

Hour 21. Laplace Transform Translation Theorems

In the previous section, we became acquainted with the Laplace transforms of some basic functions. In this section we will learn how to extend those results so that we are able to transform many more functions. Our mainstay approach will be to apply translations (shifts) to the functions we are dealing with in either the input domain or the Laplace domain.

When working with Laplace transform problems, we think of functions f(t) as belonging to an *input domain* and the Laplace transformed version F(s) as belonging to the Laplace domain. We have already seen that differentiation in the input domain manifests as multiplication by the independent variable in the Laplace domain (less the value of the original function at zero). We've seen also that multiplication by -t in the input domain manifests as differentiation with respect to s in the Laplace domain.

Multiplication of a function y(t) by $e^{\xi t}$ in the input domain has an interesting way of showing up in the Laplace transform domain: It causes the function in the Laplace domain to shift to the right along the *s* axis by the amount ξ. In other words, if the Laplace transform of f(t) is F(s), then the Laplace transform of $e^{\xi t} f(t)$ is F(s - ξ).

$$\mathcal{L}\left(e^{\xi t} f(t)\right) = \int_0^\infty e^{\xi t} f(t) e^{-st} dt$$

$$= \int_0^\infty f(t) e^{-(s-\xi)t} dt$$

$$= F(s - \xi) \qquad (21\text{-}1)$$

The insight here was to recognize, in the second integral, that s - ξ plays the role of a new Laplace parameter, i.e., $s_2 = s_1 - \xi$. The Laplace domain function F(s) is shifted to the right by the amount ξ in response to the input domain event of multiplication by $e^{\xi t}$. This is also referred to as a translation by the amount ξ.

Example 21-1: Determine the Laplace transform of $e^{3t} \sin(t)$.

Solution: Let sin(t) be f(t), a function in the input domain. Now consider the effect of multiplication by e^{3t}. It's important to keep straight that if ξ is positive, then s is replaced by s - ξ in the Laplace domain. We don't really need to *calculate* the Laplace transform all over from scratch, we simply need to replace s with s - 3 to account for the fact that sin(t) is being multiplied by e^{3t}. We have:

Differential Equations in 24 Hours

$$\sin(t) \quad \Leftrightarrow \quad \frac{1}{s^2+1}$$

$$e^{3t}\sin(t) \quad \Leftrightarrow \quad \frac{1}{(s-3)^2+1}$$

(21-2)

Thus:

$$\mathcal{L}\{e^{3t}\sin(t)\} = \frac{1}{(s-3)^2+1}$$

(21-3)

◊

The style of solution is a little different here than what we may be used to. In Eq. 21-2, we are basically saying, "If sin(t) is associated with $1/(s^2 + 1)$ then $e^{3t}\sin(t)$ is associated with $1/((s - 3)^2 + 1)$." By "associated" we mean "is a transform pair with."

Translation theorems help us quickly obtain Laplace transforms without having to go back to the original definition of the Laplace transform and crank the integral. Recall that a translation of a function Ψ(s) can be realized by replacing s with s - a. The effect is to shift Ψ(s) to the right by the amount a. Ψ(s - a) is the resulting shifted version.

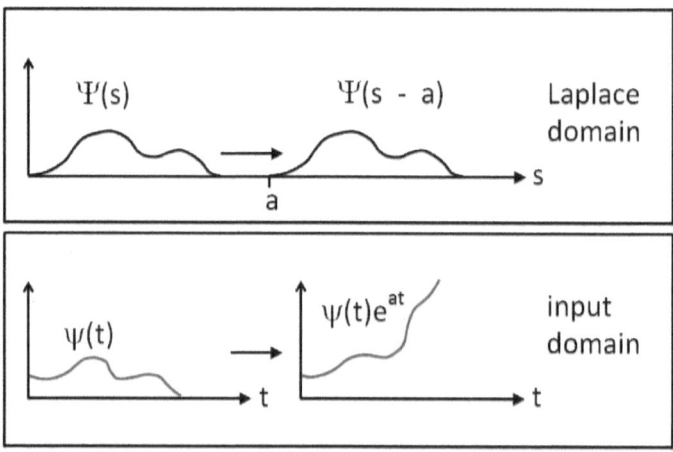

166

Hour 21. Laplace Transform Translation Theorems

But shift by the amount a in the *Laplace* domain effects multiplication by e^{at} in the input domain. Think of the two operations occurring in the two domains as happening in parallel. The first translation theorem says: If we shift by a in Laplace domain, we multiply by e^{at} in the input domain and if we multiply by e^{at} in the input domain, we shift by a in the Laplace domain.

First Translation Theorem :

If $\mathcal{L}\{f(t)\} = F(s)$ and a is any real number,

$$\mathcal{L}\{e^{at} f(t)\} = F(s - a)$$

(21-4)

●

Let's look at a couple of examples:

Example 21-2: Determine the Laplace transform of $t^{10}e^{-6t}$.

$$\mathcal{L}\{t\} = \frac{1}{s^2} \Rightarrow \mathcal{L}\{t^{10} e^{-6t}\} = \frac{10!}{(s+6)^{11}}$$

(21-5)

◊

Example 21-3: Determine the Laplace transform of $e^{2t}(t-1)^2$:

$$\begin{aligned} \mathcal{L}\{e^{2t}(t-1)^2\} &= \mathcal{L}\{t^2 e^{2t} - 2t e^{2t} + e^{2t}\} \\ &= \mathcal{L}\{t^2 e^{2t}\} - 2\mathcal{L}\{t e^{2t}\} + \mathcal{L}\{1 e^{2t}\} \\ &= \frac{2}{(s-2)^3} - \frac{2}{(s-2)^2} + \frac{1}{s-2} \end{aligned}$$

(21-6)

◊

The input domain and the Laplace domain are "two hands clapping." Thus there is an inverse theorem to accompany Eq. 21-4.

Inverse First Translation Theorem:

If $\mathcal{L}^{-1}\{F(s)\} = f(t)$ and a is any real number,

$$\mathcal{L}^{-1}\{F(s-a)\} = e^{at}f(t)$$

(21-7)

•

Let's use this theorem to see how we can go from the Laplace domain back to the input domain.

Example 21-4: Determine the inverse Laplace transform of $1/(s^2 + 2s + 5)$.

Solution: The trick here is to complete the square in the denominator.

$$\mathcal{L}^{-1}\left\{\frac{1}{s^2 + 2s + 5}\right\} = \mathcal{L}^{-1}\left\{\frac{1}{s^2 + 2s + 1 + 4}\right\}$$

$$= \mathcal{L}^{-1}\left\{\frac{1}{(s+1)^2 + 4}\right\}$$

(21-8)

Next, we can factor out ½ to put this in the sine form:

$$\mathcal{L}^{-1}\left\{\frac{1}{s^2 + 2s + 5}\right\} = \mathcal{L}^{-1}\left\{\frac{1}{(s+1)^2 + 4}\right\}$$

$$= \frac{1}{2}\mathcal{L}^{-1}\left\{\frac{2}{(s+1)^2 + 2^2}\right\}$$

$$= \frac{1}{2}e^{-t}\sin(2t)$$

(21-9)

◊

Hour 21. Laplace Transform Translation Theorems

Example 21-5: Solve $y' + y = 3e^{2t}$ with $y(0) = 0$.

Solution: Laplace transforming, we obtain:

$$sY(s) - y(0) + Y(s) = \frac{3}{s-2} \quad \text{Or:}$$

$$Y(s)(s+1) = \frac{3}{s-2} \tag{21-10}$$

Thus:

$$Y(s) = \frac{3}{(s+1)(s-2)} \tag{21-11}$$

Using partial fractions, this becomes:

$$Y(s) = \frac{-1}{(s+1)} + \frac{1}{(s-2)} \tag{21-12}$$

Now, taking the inverse Laplace transform, we have our answer:

$$y(t) = -e^{-t} + e^{2t} \tag{21-13}$$

◊

In the next example, we apply Laplace transforms to the solution of an LRC circuit with an applied voltage.

Example 21-6: LRC Circuit with zero applied voltage. The simplest possible LRC circuit consists of an inductor L, a capacitor C, and a resistor R. The current through the circuit is $j = j(t)$. The initial conditions are $j(0) = 0$ and $j'(0) = 1$. Solve for the current as a function of time.

Determining an equation to describe the circuit requires adding up the voltages across each of these elements. The current j(t) is the rate at which charge q is flowing per unit time, that is, j(t) = dq/dt. The voltage drop across the inductor is V_L = L(dj/dt). An inductor has a magnetic field in it which "pushes back" against any change in current, dj/dt. The larger the inductance, the larger the voltage drop for a given change in current. A resistor is a device that "resists" the flow of electrons. The voltage drop across the resistor is V_R = Rj(t). This reflects the fact that, the larger the resistance R, the larger the voltage needed to maintain a certain current flow through the resistor. A capacitor is a temporary charge storage device. Equal and opposite charges of amount q occur on plates separated from each other by a gap. The voltage drop across a capacitor is V_C = q/C. For a small capacitance, a large voltage is required to keep the charges separate. For a large charge, a large voltage is present. Kirchoff's Law says that these voltage drops must add up to the applied voltage which, in this case, is 0.

$$V_L + V_R + V_C = V_{APPLIED} \qquad (21\text{-}14)$$

Plugging in, we have:

$$L\frac{dj}{dt} + Rj + \frac{1}{C}q = 0 \qquad (21\text{-}15)$$

The current j is the time derivative of charge: j(t) = dq/dt. Differentiating Eq. 21-15, we obtain a second-order ordinary differential equation:

Hour 21. Laplace Transform Translation Theorems

$$L\frac{d^2j}{dt^2} + R\frac{dj}{dt} + \frac{1}{C}j = 0 \tag{21-16}$$

Remembering that R, L, and C are constants, we are now ready to apply the Laplace transform to the whole equation. Be careful not to confuse the inductance L with the Laplace transform $\mathcal{L}$:

$$\mathcal{L}\{s^2 J - s\, j(0) - j'(0)\} + R\{sJ - j(0)\} + \frac{1}{C}J = 0 \tag{21-17}$$

Plugging in our initial conditions, this becomes:

$$L s^2 J - L + R s J + \frac{1}{C} J = 0 \tag{21-18}$$

Separating variables, we have:

$$J = \frac{L}{L s^2 + R s + \frac{1}{C}} = \frac{1}{s^2 + \frac{R}{L}s + \frac{1}{LC}} \tag{21-19}$$

We can complete the square in the denominator:

$$J = \frac{1}{s^2 + \frac{R}{L}s + \frac{R^2}{4L^2} + \frac{1}{LC} - \frac{R^2}{4L^2}}$$

$$= \frac{1}{\left(s + \frac{R}{2L}\right)^2 + \frac{1}{LC} - \frac{R^2}{4L^2}} \tag{21-20}$$

If we can put this in a form where it's recognizable as a translated sine function:

$$J = \frac{1}{\sqrt{\frac{1}{LC} - \frac{R^2}{4L^2}}} \cdot \frac{\sqrt{\frac{1}{LC} - \frac{R^2}{4L^2}}}{\left\{\left(s + \frac{R}{2L}\right)^2 + \frac{1}{LC} - \frac{R^2}{4L^2}\right\}}$$

(21-21)

Thus:

$$j(t) = \frac{1}{\sqrt{\frac{1}{LC} - \frac{R^2}{4L^2}}} e^{-\frac{R}{2L}t} \sin\left(\sqrt{\frac{1}{LC} - \frac{R^2}{4L^2}}\, t\right)$$

(21-22)

It may seem like a bit of a leap from Eq. 21-21 to Eq. 21-22; however, if we look at the components, it's straightforward. The right-most term is in the form $\frac{a}{\left(s - \frac{R}{2L}\right)^2 + a^2}$ and the translation R/2L in the Laplace domain was accounted for in the input domain as multiplication by $e^{-(R/2L)t}$. Thus we have sin(at) with $e^{-(R/2L)t}$ and a constant term in front.

Eq. 21-22 describes a current that oscillates but decays over time so long as the resistance R is greater than zero and less than $2(L/C)^{1/2}$. If R = 0, the solution reduces to the harmonic oscillator solution for which the oscillations do not decay:

If R = 0.

$$j(t) = \sqrt{LC} \sin\left(\frac{1}{\sqrt{LC}} t\right)$$

(21-23)

◊

Hour 21. Laplace Transform Translation Theorems

If you listen to the din of system engineers discussing digital signal processing, you are likely to hear the word "impulse response" being uttered. What engineers are referring to when they speak about impulse response is quite literally the *response* of a system when an *impulse* function is applied to the system. For a broad class of systems, how the system responds when a narrow impulse function is applied, provides a comprehensive characterization of all of the system's attributes. In particular, for linear systems, if the response of the system to an impulse is known, then the response of that system to any input can be predicted.

When dealing with functions of a continuous variable, an object known as a Dirac delta function is used to represent an impulse. The attributes of a Dirac delta function are these: 1.) It has constant area; 2.) It has infinite height; 3.) It has zero width; and 4.) It is centered over the value that makes its argument zero.

We can construct a Dirac delta function from the step function we worked with in Hour 19. Recall that we discussed the following step function:

Consider the function:

$$f(t) = \begin{cases} 0, & 0 \leq t < a \\ b, & a \leq t < \infty \end{cases} \quad (21\text{-}24)$$

It can be represented as $b\, u(t - a)$ and appears as follows:

173

Differential Equations in 24 Hours

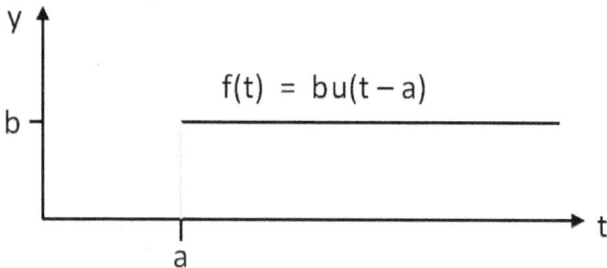

As we have already seen, the Laplace transform of this function is:

$$\mathcal{L}(f(t)) = \int_0^a 0\, dt + \int_a^\infty b e^{-st} dt = \frac{b}{s} e^{-as}$$

$$F(s) = \frac{b}{s} e^{-as} \qquad (21\text{-}25)$$

If we let a = t_0 and b = 1/h then the step, which we will call $y_1(t)$, has the following appearance:

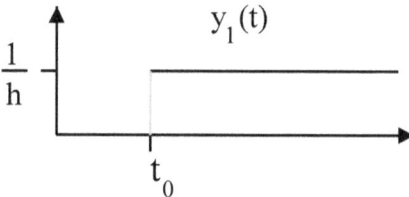

Plugging into Eq. 21-25, the Laplace transform of $y_1(t)$ is:

$$Y_1(s) = \frac{1}{sh} e^{-t_0 s} \qquad (21\text{-}26)$$

We can make another step function, $y_2(t)$, shifted over by the amount h and having the same height 1/h:

174

Hour 21. Laplace Transform Translation Theorems

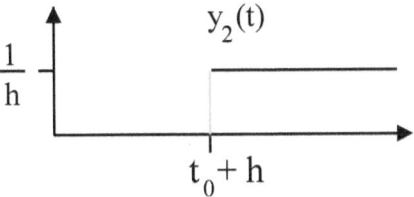

Again plugging into Eq. 21-25, the Laplace transform of this is:

$$Y_2(s) = \frac{1}{sh} e^{-(t_0 + h)s} \tag{21-27}$$

We can form an impulse of unit area, width h, and height 1/h by forming y(t) = $y_1(t) - y_2(t)$. Behold:

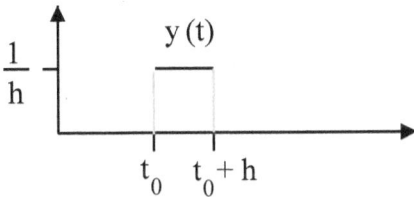

Since the Laplace transform is a linear transform, we can determine the Laplace transform of y(t) easily since we know the Laplace transforms of $y_1(t)$ and $y_2(t)$.

$$Y(s) = \mathcal{L}\{y_1(t) - y_2(t)\} = \mathcal{L}\{y_1(t)\} - \mathcal{L}\{y_2(t)\}$$
$$= Y_1(s) - Y_2(s) = \frac{1}{sh} e^{-t_0 s} - \frac{1}{sh} e^{-(t_0 + h)s}$$
$$= \frac{1}{sh} e^{-t_0 s} \left(1 - e^{-hs}\right) = e^{-t_0 s} \left\{ \frac{1 - e^{-hs}}{hs} \right\}$$

$$(21-28)$$

The Dirac delta function is the limiting case of this function y(t) for infinite height and zero width. This occurs as h approaches zero:

Differential Equations in 24 Hours

Dirac delta function:

$$\delta(t - t_0) = \lim_{h \to 0} y(t) \quad (21\text{-}29)$$

Again, the spike of the Dirac delta function occurs at $t = t_0$ and this is the value of t that makes the argument of the Dirac delta function zero. The Laplace transform of the Dirac delta function can be approached by taking the Laplace transform of Eq. 21-29 and then letting h again approach zero.

$$\mathcal{L}\{\delta(t - t_0)\} = \lim_{h \to 0} e^{-t_0 s} \left\{ \frac{1 - e^{-hs}}{hs} \right\}$$

$$= e^{-t_0 s} \lim_{h \to 0} \left\{ \frac{1 - e^{-hs}}{hs} \right\} \quad (21\text{-}30)$$

Using l'Hopital's rule, being careful to differentiate with respect to s, this limit can be evaluated:

$$\mathcal{L}\{\delta(t - t_0)\} = e^{-t_0 s} \lim_{h \to 0} \left\{ \frac{h\, e^{-hs}}{h} \right\} = 1\, e^{-t_0 s}$$

(21-31)

We are now in a position to extend our concept of a derivative. In particular, we are now able to take the derivative of a step.

Example 21-7: Derivative of a step. Take the derivative of the following step function:

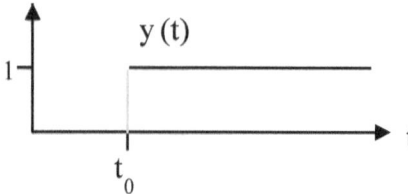

Solution:

The step appears in the Laplace domain as:

Hour 21. Laplace Transform Translation Theorems

$$\mathcal{L}\left\{\frac{dy}{dt}\right\} = s\mathcal{L}(y) - y(0) = s\frac{1}{s}e^{-t_0 s} - 0 = e^{-t_0 s}$$

(21-32)

But this is the Laplace transform of the Dirac delta function centered at t_0. Thus:

$$\mathcal{L}\left\{\frac{dy}{dt}\right\} = \mathcal{L}\{\delta(t - t_0)\} \qquad (21\text{-}33)$$

◊

There is a second translation theorem for Laplace transforms. If a function F(s) in the Laplace domain is multiplied by e^{-as}, the effect in the input domain is twofold: the function f(t) is shifted to the right by the amount a and then this shifted function is set equal to zero for any times less than a. That is, f(t) is replaced by f(t - a)u(t - a):

Second Translation Theorem :

If $F(s) = \mathcal{L}\{f(t)\}$ and $a > 0$, then

$$\mathcal{L}\{f(t - a)u(t - a)\} = e^{-as} F(s) \qquad (21\text{-}34)$$

•

Thus multiplication by e^{-as} has the following appearance in the time domain:

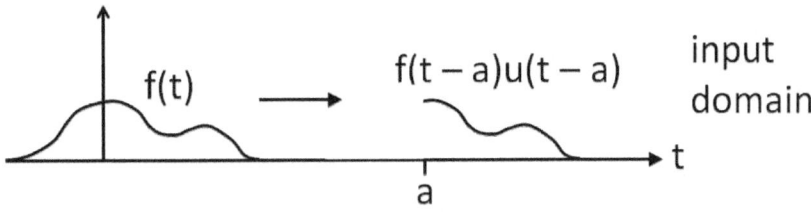

Example 21-8: Struck Harmonic Oscillator. Solve $y'' + 16y = \delta(t - 2\pi)$ with initial conditions $y(0) = 0$ and $y'(0) = 0$.

Solution: This is a system that is at rest until it is struck by an impulse at time $t = 2\pi$. If we transform the equation, we have:

$$s^2 Y(s) - s y(0) - y'(0) + 16Y(s) = e^{-2\pi s}$$

(21-35)

The initial conditions are both zero; therefore, this reduces to:

$$(16 + s^2)Y(s) = e^{-2\pi s}$$

(21-36)

Or:

$$Y(s) = \frac{1}{4} \frac{4e^{-2\pi s}}{(s^2 + 16)}$$

(21-37)

Now this is in the form that sine takes when transformed to the Laplace domain. In this case, it is shifted in the input domain by 2π (and truncated for time $t < 2\pi$) and has frequency $a = 4$. So, transforming back to the input domain, the solution is:

$$y = \frac{1}{4}\sin(4(t - 2\pi))u(t - 2\pi) = \frac{1}{4}\sin(4t)u(t - 2\pi)$$

(21-38)

Or:

$$y = \begin{cases} 0, & t < 2\pi \\ \frac{1}{4}\sin(4t), & t \geq 2\pi \end{cases}$$

(21-39)

Hour 21. Laplace Transform Translation Theorems

We can illustrate the solution as follows:

The system is quiet until it is hit with an impulse at time $t = 2\pi$ at which time it oscillates with amplitude ¼.

◊

Let's understand why the Second Translation Theorem makes sense. Consider the Laplace transform of $f(t-a)u(t-a)$ and let's show how it gives $e^{-as}F(s)$. We multiply by 1 in the form $e^{-as}e^{as}$ and bring the e^{as} part into the integral. Then we make a special change of variables. We have:

$$\int_0^\infty f(t-a)u(t-a)e^{-st}dt = e^{-as}e^{as}\int_0^\infty f(t-a)u(t-a)e^{-st}dt =$$

$$e^{-as}\int_0^\infty f(t-a)u(t-a)e^{-s(t-a)}dt = e^{-as}\int_a^\infty f(t-a)e^{-s(t-a)}d(t-a)$$

(21-40)

Now introduce a change of variables, letting $\tau = t - a$ we have:

$$e^{-as}\int_a^\infty f(t-a)e^{-s(t-a)}d(t-a) = e^{-as}\int_0^\infty f(\tau)e^{-s\tau}d\tau$$

Differential Equations in 24 Hours

(21-41)

$$= e^{-as}F(s)$$

(21-42)

Which is what we wanted to show.

Hour 21. Laplace Transform Translation Theorems

Paul A. M. Dirac
(1902-1984)
Illustration © 2015, Brandon Ross Sugg

Paul Dirac was born in England on August 8, 1902. In 1923 he received a BA in mathematics from University of Bristol after completing a BS in engineering two years earlier. He received a PhD in physics from Cambridge in 1926.

Dirac discovered the relativistic equation for the electron, an equation which bears his name. This equation predicts the existence of antimatter. Dirac introduced his delta function in his 1930 book, The Principles of Quantum Mechanics. Dirac received a Nobel Prize for physics in 1933. Much of quantum mechanics is written in Dirac notation. Dirac notation for the inner product between two functions f and h is $<f \mid h>$, that is:

$$<f \mid h> = \int f(t) h^*(t) \, dt \qquad (21\text{-}43)$$

Where $h^*(t)$ is the complex conjugate of $h(t)$.

Dirac was a man of few words. At Cambridge, his colleagues defined a unit of speech as a "Dirac." A Dirac is one word per hour.

Hour 22. Laplace Transform Translation Theorems Exercises

Exercise 22-1: Use Laplace transforms to solve the initial value problem $y'' + 4y' + 13y = 0$ with initial conditions $y(0) = 1$ and $y'(0) = -2$.

Exercise 22-2: Use Laplace transforms to solve the initial value problem $\ddot{y} + y = e^t$ with initial conditions $y(0) = 1$ and $\dot{y}(0) = 1$.

Exercise 22-3: Solve, using Laplace transforms, the initial value problem:

$y'' + 5y' + 6y = 0$ with initial conditions $y(0) = 0$ and $y'(0) = 1$.

Exercise 22-4: Solve the initial value problem: $\ddot{y} + 9y = \delta(t - 3\pi/2)$ with initial conditions $y(0) = 0$ and $y'(0) = 0$.

Exercise 22-5: An LRC circuit has an applied voltage $e^{-\omega t}$. We wish to solve for the current $j = j(t)$. The initial conditions are $j(0) = 0$ and $j'(0) = 0$.

Applying Kirchoff's Law, the voltages add up as follows:

$$L\frac{dj}{dt} + Rj + \frac{1}{C}Q = e^{-\omega t} \qquad (22\text{-}1)$$

The current j is the time derivative of charge: $j(t) = dQ/dt$. Differentiating Eq. 22-1, we obtain a second-order ordinary differential equation:

183

Differential Equations in 24 Hours

$$L\frac{d^2 j}{dt^2} + R\frac{dj}{dt} + \frac{1}{C}j = -\omega e^{-\omega t} \tag{22-2}$$

Show that the Laplace transform of the current, J(s) is:

$$J(s) = -\frac{\omega}{(\omega + s)} \frac{1}{(Ls^2 + Rs + \frac{1}{C})} \tag{22-3}$$

Exercise 22-6: For the previous problem, solve for j = j(t) if ω = 1, L = 1, R = 2, C = ½.

Part four:

Introduction to Systems of Differential Equations

Hour 23.　　First-Order Linear Systems with Constant Coefficients.......187

Hour 24.　　First-Order Linear Systems Exercises...................................199

Hour 23. First-Order Linear Systems with Constant Coefficients

A *system* is a set of equations which are true simultaneously. In this section we will see how some simple systems of differential equations are solved. We group the equations with a single brace on the left to show that they are intimately connected to one another. In this section, a prime ' will denote differentiation with respect to time.

A system of n first order linear differential equations has the form:

$$\begin{cases} y_1' = p_{11}(t)y_1 + p_{12}(t)y_2 + \ldots + p_{1n}(t)y_n + g_1(t) \\ y_2' = p_{21}(t)y_1 + p_{22}(t)y_2 + \ldots + p_{2n}(t)y_n + g_2(t) \\ \vdots \quad\quad \vdots \quad\quad\quad \vdots \quad\quad\quad\quad\quad \vdots \quad\quad\quad \vdots \\ y_n' = p_{n1}(t)y_1 + p_{n2}(t)y_2 + \ldots + p_{nn}(t)y_n + g_n(t) \end{cases} \tag{23-1}$$

A *constant coefficient homogeneous system* has $g_1(t) = g_2(t) = \ldots = g_n(t) = 0$ and all the p's are constants. These kinds of systems can be solved by reducing the system to an **eigenvalue equation**, an equation involving **eigenvectors** and **eigenvalues**. Once the eigenvalues and eigenvectors are determined, the solution can be arrived at using a few simple matrix operations. The best way to learn this is to see an example.

Example 23-1: The following is a constant coefficient homogeneous system:

$$\begin{cases} y_1' = y_1 + 2y_2 \\ y_2' = 2y_1 + y_2 \end{cases} \tag{23-2}$$

This can be written:

$$\vec{y}' = A\vec{y} \tag{23-3}$$

Where:

$$\vec{y} = \begin{bmatrix} y_1 \\ y_2 \end{bmatrix}, \quad A = \begin{bmatrix} 1 & 2 \\ 2 & 1 \end{bmatrix}$$

Differential Equations in 24 Hours

The trick to solving a system like this is to look for solutions of the form:

$$\vec{y} = e^{\lambda t}\vec{x} \qquad (23\text{-}4)$$

Where the components of $\vec{x}$ are constants. Plugging into 23-3, we have:

$$(e^{\lambda t}\vec{x})' = Ae^{\lambda t}\vec{x} \qquad (23\text{-}5)$$

Which becomes:

$$\lambda e^{\lambda t}\vec{x} + e^{\lambda t}\vec{x}' = e^{\lambda t}A\vec{x} \qquad (23\text{-}6)$$

$\vec{x}$ has constant components; therefore, $\vec{x}'$ is zero $\vec{0}$. Then 23-6 becomes:

$$A\vec{x} = \lambda\vec{x} \qquad (23\text{-}7)$$

This is an *eigenvalue equation*. λ is an *eigenvalue* and $\vec{x}$ is an *eigenvector*.

We can use linear algebra to find the eigenvalues and eigenvectors. We introduce the identity matrix into 23-7:

$$A\vec{x} = \lambda I\vec{x} \qquad (23\text{-}8)$$

Or:

$$(A - \lambda I)\vec{x} = \vec{0} \qquad (23\text{-}9)$$

Alternately we can write:

$$(\lambda I - A)\vec{x} = \vec{0} \qquad (23\text{-}10)$$

To have a non-trivial solution, this requires:

$$\det(A - \lambda I) = 0 \qquad (23\text{-}11)$$

Hour 23. First-Order Linear Systems with Constant Coefficients

Or:

$$\det(\lambda I - A) = \begin{vmatrix} (\lambda-1) & -2 \\ -2 & (\lambda-1) \end{vmatrix} = 0 \quad (23\text{-}12)$$

Thus:

$$(\lambda - 1)^2 - 4 = 0 \quad (23\text{-}13)$$

Or:

$$\lambda^2 - 2\lambda - 3 = 0 \quad (23\text{-}14)$$

Or:

$$(\lambda + 1)(\lambda - 3) = 0 \quad (23\text{-}15)$$

Thus we have two eigenvalues: $\lambda_1 = -1$, $\lambda_2 = 3$.

We can solve for the eigenvector $\vec{x}_1$ associated with $\lambda_1 = -1$. We have:

$$(\lambda_1 I - A)\vec{x}_1 = \vec{0} \quad (23\text{-}16)$$

Which we can write as:

$$\begin{bmatrix} (\lambda_1 - 1) & -2 \\ -2 & (\lambda_1 - 1) \end{bmatrix} \begin{bmatrix} x_{11} \\ x_{12} \end{bmatrix} = \begin{bmatrix} 0 \\ 0 \end{bmatrix} \quad (23\text{-}17)$$

Or:

$$\begin{bmatrix} -2 & -2 \\ -2 & -2 \end{bmatrix} \begin{bmatrix} x_{11} \\ x_{12} \end{bmatrix} = \begin{bmatrix} 0 \\ 0 \end{bmatrix} \quad (23\text{-}18)$$

Or:

$$\begin{bmatrix} -2 & -2 & 0 \\ -2 & -2 & 0 \end{bmatrix} \quad (23\text{-}19)$$

Using elementary row operations, we have:

$$\begin{bmatrix} 1 & 1 & 0 \\ 0 & 0 & 0 \end{bmatrix} \quad (23\text{-}20)$$

Thus $x_{11} = -x_{12}$. Let $x_{11} = s$. Then $x_{12} = -s$. Therefore:

$$\vec{x}_1 = s \begin{bmatrix} 1 \\ -1 \end{bmatrix} \quad (23\text{-}21)$$

s cancels out of each side of:

$$As \begin{bmatrix} 1 \\ -1 \end{bmatrix} = \lambda_1 s \begin{bmatrix} 1 \\ -1 \end{bmatrix} \quad (23\text{-}22)$$

So we are justified in writing the eigenvectors without the s:

$$\vec{x}_1 = \begin{bmatrix} 1 \\ -1 \end{bmatrix} \quad (23\text{-}23)$$

Now the eigenvector $\vec{x}_2$ associated with $\lambda_2 = 3$ satisfies:

$$(\lambda_2 I - A)\vec{x}_2 = \vec{0} \quad (23\text{-}24)$$

Which is:

$$\begin{bmatrix} (\lambda_2 - 1) & -2 \\ -2 & (\lambda_2 - 1) \end{bmatrix} \begin{bmatrix} x_{21} \\ x_{22} \end{bmatrix} = \begin{bmatrix} 0 \\ 0 \end{bmatrix} \quad (23\text{-}25)$$

Hour 23. First-Order Linear Systems with Constant Coefficients

Which becomes:

$$\begin{bmatrix} 2 & -2 \\ -2 & 2 \end{bmatrix} \begin{bmatrix} x_{21} \\ x_{22} \end{bmatrix} = \begin{bmatrix} 0 \\ 0 \end{bmatrix} \qquad (23\text{-}26)$$

Or:

$$\begin{bmatrix} 2 & -2 & 0 \\ -2 & 2 & 0 \end{bmatrix} \qquad (23\text{-}27)$$

Using elementary row operations, this becomes:

$$\begin{bmatrix} 1 & -1 & 0 \\ 0 & 0 & 0 \end{bmatrix} \qquad (23\text{-}28)$$

Thus $x_{21} = x_{22}$. Let $x_{21} = u$. Then $x_{22} = u$. Thus:

$$\vec{x}_2 = u \begin{bmatrix} 1 \\ 1 \end{bmatrix} \qquad (23\text{-}29)$$

Plugging in, we have:

$$Au \begin{bmatrix} 1 \\ 1 \end{bmatrix} = \lambda_2 u \begin{bmatrix} 1 \\ 1 \end{bmatrix} \qquad (23\text{-}30)$$

u cancels out; therefore, we are justified in simply writing:

$$\vec{x}_2 = \begin{bmatrix} 1 \\ 1 \end{bmatrix} \qquad (23\text{-}31)$$

So the solution of 23-2 is the solution of:

$$\vec{y}' = A\vec{y} \qquad (23\text{-}32)$$

Which is:

$$\vec{y} = c_1\vec{q}_1 + c_2\vec{q}_2 = c_1 e^{\lambda_1 t}\vec{x}_1 + c_2 e^{\lambda_2 t}\vec{x}_2 \quad (23\text{-}33)$$

$$= c_1 e^{-t}\begin{bmatrix}1\\-1\end{bmatrix} + c_2 e^{3t}\begin{bmatrix}1\\1\end{bmatrix} \quad (23\text{-}34)$$

$$= \begin{bmatrix}c_1 e^{-t}\\-c_1 e^{-t}\end{bmatrix} + \begin{bmatrix}c_2 e^{3t}\\c_2 e^{3t}\end{bmatrix} \quad (23\text{-}35)$$

Add vectors by adding their components:

$$\vec{y} = \begin{bmatrix}c_1 e^{-t} + c_2 e^{3t}\\-c_1 e^{-t} + c_2 e^{3t}\end{bmatrix} \quad (23\text{-}36)$$

Or:

$$\vec{y} = \begin{bmatrix}e^{-t} + e^{3t}\\-e^{-t} + e^{3t}\end{bmatrix}\begin{bmatrix}c_1\\c_2\end{bmatrix} \quad (23\text{-}37)$$

Equation 23-37 is the general solution because $\det(\vec{q}_1, \vec{q}_2)$ does not vanish:

$$\begin{vmatrix}e^{-t} & e^{3t}\\-e^{-t} & e^{3t}\end{vmatrix} = 2e^{2t} \neq 0 \quad (23\text{-}38)$$

◊

In case of a complex eigenvalue λ and eigenvector $\vec{x} = \vec{v} + i\vec{w}$, we can form:

$$\vec{y} = e^{\lambda t}(\vec{v} + i\vec{w}) \quad (23\text{-}39)$$

Hour 23. First-Order Linear Systems with Constant Coefficients

And use Euler's formula to expand $e^{\lambda t}$. Then:

$$e^{\lambda t}(\vec{v} + i\vec{w}) = \vec{q}_1 + i\vec{q}_2 \qquad (23\text{-}40)$$

The general solution then is:

$$\vec{y}_g = c_1\vec{q}_1 + c_2\vec{q}_2 \qquad (23\text{-}41)$$

Provided that:

$$\det([\vec{q}_1, \vec{q}_2]) \neq 0 \qquad (23\text{-}42)$$

Example 23-2: Constant coefficient homogeneous system with complex eigenvalues:

$$\begin{cases} y_1' = y_1 + y_2 \\ y_2' = -y_1 + y_2 \end{cases} \qquad (23\text{-}43)$$

This can be written in the form of an eigenvalue equation.

$$\vec{y}' = A\vec{y} \qquad (23\text{-}44)$$

Here we have:

$$A = \begin{bmatrix} 1 & 1 \\ -1 & 1 \end{bmatrix} \qquad (23\text{-}45)$$

We introduce the identity matrix into 23-7:

$$A\vec{x} = \lambda I \vec{x} \qquad (23\text{-}46)$$

Or:

$$(A - \lambda I)\vec{x} = \vec{0} \qquad (23\text{-}47)$$

Differential Equations in 24 Hours

To have a non-trivial solution, this requires:

$$\det(A - \lambda I) = 0 \qquad (23\text{-}48)$$

We have:

$$\lambda I - A = \begin{bmatrix} (\lambda - 1) & -1 \\ 1 & (\lambda - 1) \end{bmatrix} \qquad (23\text{-}49)$$

So that:

$$\det(\lambda I - A) = (\lambda - 1)^2 + 1 = 0 \qquad (23\text{-}50)$$

$$(\lambda - 1)^2 + 1 = 0 \qquad (23\text{-}51)$$

Or:

$$\lambda^2 - 2\lambda + 2 = 0 \qquad (23\text{-}52)$$

Thus:

$$\lambda = \frac{2 \pm \sqrt{4 - 8}}{2} = 1 \pm i \qquad (23\text{-}53)$$

Or:

$$\begin{aligned} \lambda_1 &= 1 + i \\ \lambda_2 &= 1 - i \end{aligned} \qquad (23\text{-}54)$$

Working with the first eigenvalue $\lambda_1 = 1 + i$, and using 23-49 in 23-10, we have:

$$\begin{bmatrix} i & -1 \\ 1 & i \end{bmatrix} \begin{bmatrix} x_{11} \\ x_{12} \end{bmatrix} = \begin{bmatrix} 0 \\ 0 \end{bmatrix} \qquad (23\text{-}55)$$

Hour 23. First-Order Linear Systems with Constant Coefficients

Here x_{11} and x_{12} are the first and second components of eigenvector $\vec{x}_1$. 23-55 may be written:

$$\begin{bmatrix} i & -1 & 0 \\ 1 & i & 0 \end{bmatrix} \quad (23\text{-}56)$$

Using elementary row operations, this becomes:

$$\begin{bmatrix} 1 & i & 0 \\ 0 & 0 & 0 \end{bmatrix} \quad (23\text{-}57)$$

Which means that:

$$x_{11} = -i x_{12} \quad (23\text{-}58)$$

Let $x_{11} = s$. Then $x_{12} = is$. Thus the eigenvector $\vec{x}_1$ is:

$$\vec{x}_1 = s \begin{bmatrix} 1 \\ i \end{bmatrix} \quad (23\text{-}59)$$

But s cancels out of 23-7; therefore we are justified in writing: uniqueness

$$\vec{x}_1 = \begin{bmatrix} 1 \\ i \end{bmatrix} \quad (23\text{-}60)$$

Note that we can write:

$$\begin{bmatrix} 1 \\ i \end{bmatrix} = \vec{x}_1 = \vec{v}_1 + i\vec{w}_1 = \begin{bmatrix} 1 \\ 0 \end{bmatrix} + i \begin{bmatrix} 0 \\ 1 \end{bmatrix} \quad (23\text{-}61)$$

Plugging into 23-39, we have:

Differential Equations in 24 Hours

$$\vec{y} = e^{(1+i)t}\vec{x}_1 = e^t e^{it}\vec{x}_1 = e^t(\cos t + i\sin t)\vec{x}_1$$
$$= e^t(\cos t + i\sin t)\left(\begin{bmatrix}1\\0\end{bmatrix} + i\begin{bmatrix}0\\1\end{bmatrix}\right) \quad (23\text{-}62)$$

Expanding and separating real and imaginary parts, we have:

$$\vec{y} = \begin{bmatrix}e^t\cos t\\0\end{bmatrix} + i\begin{bmatrix}0\\e^t\cos t\end{bmatrix} + i\begin{bmatrix}e^t\sin t\\0\end{bmatrix} - \begin{bmatrix}0\\e^t\sin t\end{bmatrix}$$
$$= \begin{bmatrix}e^t\cos t\\-e^t\sin t\end{bmatrix} + i\begin{bmatrix}e^t\sin t\\e^t\cos t\end{bmatrix} \quad (23\text{-}63)$$

Referring back to 23-40, we can write:

$$\vec{y} = \vec{q}_1 + i\vec{q}_2 = \begin{bmatrix}e^t\cos t\\-e^t\sin t\end{bmatrix} + i\begin{bmatrix}e^t\sin t\\e^t\cos t\end{bmatrix} \quad (23\text{-}64)$$

Plugging into 23-41, we have:

$$\vec{y}_g = c_1\vec{q}_1 + c_2\vec{q}_2 = c_1\begin{bmatrix}e^t\cos t\\-e^t\sin t\end{bmatrix} + c_2\begin{bmatrix}e^t\sin t\\e^t\cos t\end{bmatrix}$$
$$= \begin{bmatrix}e^t\cos t & e^t\sin t\\-e^t\sin t & e^t\cos t\end{bmatrix}\begin{bmatrix}c_1\\c_2\end{bmatrix} \quad (23\text{-}65)$$

Which is the solution we have been working towards.

It may seem surprising that we completely solved the problem using only the first eigenvalue and eigenvector. What if we had used the other eigenvalue instead? Would we get the same solution? From 23-54, $\lambda_2 = 1 - i$. Plugging 23-49 into 23-10, we have:

$$\begin{bmatrix}-i & -1\\1 & -i\end{bmatrix}\begin{bmatrix}x_{21}\\x_{22}\end{bmatrix} = \begin{bmatrix}0\\0\end{bmatrix} \quad (23\text{-}66)$$

Hour 23. First-Order Linear Systems with Constant Coefficients

This may be written:

$$\begin{bmatrix} -i & -1 & 0 \\ 1 & -i & 0 \end{bmatrix} \quad (23\text{-}67)$$

Using elementary row operations, we have:

$$\begin{bmatrix} 1 & -i & 0 \\ 0 & 0 & 0 \end{bmatrix} \quad (23\text{-}68)$$

Which gives us:

$$x_{21} = i x_{22} \quad (23\text{-}69)$$

Letting $x_{21} = s$, we have $x_{22} = -is$. The eigenvector then is:

$$\vec{x}_2 = \begin{bmatrix} 1 \\ -i \end{bmatrix} \quad (23\text{-}70)$$

So that:

$$\begin{aligned}
\vec{y} &= e^{(1-i)t}\begin{bmatrix} 1 \\ -i \end{bmatrix} = (e^t \cos t - ie^t \sin t)\left(\begin{bmatrix} 1 \\ 0 \end{bmatrix} + i\begin{bmatrix} 0 \\ -1 \end{bmatrix}\right) \\
&= e^t \cos t \begin{bmatrix} 1 \\ 0 \end{bmatrix} + ie^t \cos t \begin{bmatrix} 0 \\ -1 \end{bmatrix} - ie^t \sin t \begin{bmatrix} 1 \\ 0 \end{bmatrix} + e^t \sin t \begin{bmatrix} 0 \\ -1 \end{bmatrix} \\
&= \begin{bmatrix} e^t \cos t \\ -e^t \sin t \end{bmatrix} - i\begin{bmatrix} e^t \sin t \\ e^t \cos t \end{bmatrix}
\end{aligned}$$

$$(23\text{-}71)$$

The minus sign can be represented in the second constant when we write:

$$\vec{y}_g = c_1 \begin{bmatrix} e^t \cos t \\ -e^t \sin t \end{bmatrix} + c_2 \begin{bmatrix} e^t \sin t \\ e^t \cos t \end{bmatrix} \qquad (23\text{-}72)$$

Which is the same as the result we obtained using the first eigenvector.

◊

Hour 24. First-Order Linear Systems Exercises

Exercise 24-1: Solve the system of equations:

$$\frac{dy_1}{dt} = y_1 + y_2$$
$$\frac{dy_2}{dt} = 4y_1 - 2y_2$$

Exercise 24-2: Solve the system of equations:

$$\frac{dy_1}{dt} = 4y_1 - 2y_2$$
$$\frac{dy_2}{dt} = 5y_1 + 2y_2$$

Exercise 24-3: Solve the system of equations:

$$\frac{dy_1}{dt} = -y_1 - y_2$$
$$\frac{dy_2}{dt} = 6y_1 + 4y_2$$

With initial conditions $y_1(0) = -2$, $y_2(0) = 6$.

Exercise 24-4: Solve the system of equations initial value problem:

$$\vec{y}' = \begin{bmatrix} -5 & -2 \\ 12 & 5 \end{bmatrix} \vec{y} \;,\; \vec{y}(1) = \begin{bmatrix} 1 \\ 0 \end{bmatrix}$$

Exercise 24-5: Solve the system of equations initial value problem:

$$\vec{y}' = \begin{bmatrix} 1 & 2 \\ 2 & 1 \end{bmatrix} \vec{y} \;,\; \vec{y}(-1) = \begin{bmatrix} 2 \\ 2 \end{bmatrix}$$

Appendix 1: Answers to Exercises

Exercise 1-1:

Let $r = u$ and $q = v^{-1}$. Then:

$$\frac{d}{dx}(uv^{-1}) = v^{-1}\frac{du}{dx} + u\frac{d}{dx}(v^{-1}) = \frac{v}{v^2}\frac{du}{dx} + u(-v^{-2})\frac{dv}{dx}$$

$$= \frac{v\frac{du}{dx} - u\frac{dv}{dx}}{v^2}$$

Exercise 1-2:

Let $u = \sin x$ and $v = \cos x$. Then:

$$\frac{v\frac{du}{dx} - u\frac{dv}{dx}}{v^2} = \frac{\cos x \cos x - \sin x(-\sin x)}{\cos^2 x} = \frac{1}{\cos^2 x} = \sec^2 x$$

Exercise 1-3:

$$\frac{d}{dx}(\sec x) = \frac{d}{dx}(\cos x)^{-1} = -(\cos x)^{-2}(-\sin x) = \frac{1}{\cos x}\frac{\sin x}{\cos x}$$

$$= \sec x \tan x$$

Exercise 1-4:

$$\int \sec x \, dx = \int \sec x \, \frac{\sec x + \tan x}{\sec x + \tan x} dx$$

$$= \int \frac{1}{\sec x + \tan x}(\sec x \tan x + \sec^2 x)dx = \ln(\sec x + \tan x) + C$$

Differential Equations in 24 Hours

Exercise 2-1:

$$y = y_0 e^{-t\frac{\ln 2}{t_{half}}}$$

$$\therefore y' = y_0 e^{-t\frac{\ln 2}{t_{half}}} \left(-\frac{\ln 2}{t_{half}}\right)$$

$$\therefore y + \left(\frac{t_{half}}{\ln 2}\right) y' = y_0 e^{-t\frac{\ln 2}{t_{half}}} + \left(-\frac{\ln 2}{t_{half}}\right)\left(\frac{t_{half}}{\ln 2}\right)\left(y_0 e^{-t\frac{\ln 2}{t_{half}}}\right) = 0$$

$$\therefore y = y_0 e^{-t\frac{\ln 2}{t_{half}}} \text{ is a solution of } y + \left(\frac{t_{half}}{\ln 2}\right) y' = 0$$

Exercise 2-2:

$$y = C_1 \cos x + C_2 \sin x$$

$$\therefore y' = -C_1 \sin x + C_2 \cos x$$

$$\therefore y'' = -C_1 \cos x - C_2 \sin x = -y$$

$$\therefore y'' + y = 0$$

Exercise 2-3:

$$y = y_0 \sqrt{1 - \frac{v^2}{c^2}} \therefore y' = y_0 \cdot \frac{1}{2}\left(1 - \frac{v^2}{c^2}\right)^{-\frac{1}{2}} \cdot \left(-\frac{2v}{c^2}\right)$$

$$\therefore \text{when } k = \frac{c^2}{y_0^2}$$

$$v + kyy' = v + \frac{c^2}{y_0^2} \cdot y_0 \sqrt{1 - \frac{v^2}{c^2}} \cdot y_0 \cdot \frac{1}{2}\left(1 - \frac{v^2}{c^2}\right)^{-\frac{1}{2}} \cdot \left(-\frac{2v}{c^2}\right) = v - v = 0$$

Appendix 1: Answers to Exercises

$\therefore y = y_0 \sqrt{1 - \frac{v^2}{c^2}}$ is a solution of $v + kyy' = 0$

Exercise 2-4:

$y(t) = e^{k\cos 2t}$

$\therefore y' = k(-2\sin 2t)e^{k\cos 2t}$ Since $y' + \sin 2t\, y = 0$

$\therefore k(-2\sin 2t)e^{k\cos 2t} + \sin 2t\, e^{k\cos 2t} = 0$

$\therefore k = \frac{1}{2}$

Exercise 3-1:

$y + b\frac{dy}{dt} = 0$

$\therefore y\, dt + b\, dy = 0$

$\therefore dt + b \cdot \frac{1}{y} dy = 0$

$\therefore t + b\ln(y) = \ln C$

$\therefore \ln y^b = \ln \frac{C}{e^t}$

$\therefore y = \sqrt[b]{\frac{C}{e^t}} = C \cdot e^{-\frac{t}{b}}$

Exercise 3-2:

$\frac{dy}{dx} = y^2 - 4 = (y+2)(y-2)$

Differential Equations in 24 Hours

$$\therefore \frac{dy}{(y+2)(y-2)} - dx = 0$$

Let $\dfrac{dy}{(y+2)(y-2)} = \dfrac{A}{y+2} + \dfrac{B}{y-2}$

$\therefore Ay - 2A + By + 2B = 1$

$\therefore A + B = 0 \quad 2B - 2A = 1$

$\therefore A = -\dfrac{1}{4} \quad B = \dfrac{1}{4}$

$\therefore \dfrac{-\frac{1}{4}}{y+2} dy + \dfrac{\frac{1}{4}}{y-2} dy - dx = 0$

$\therefore \dfrac{1}{4}\ln(y-2) - \dfrac{1}{4}\ln(y+2) - x = \ln C$

$\therefore \ln\left(\dfrac{y-2}{y+2}\right)^{\frac{1}{4}} = \ln C \cdot e^x$

$\therefore \left(\dfrac{y-2}{y+2}\right)^{\frac{1}{4}} = C \cdot e^x$

$\therefore \dfrac{y-2}{y+2} = C \cdot e^{4x}$ Let $\dfrac{y-2}{y+2} = K$

$\therefore y = \dfrac{2K+2}{1-K}$

$\therefore y = \dfrac{2C \cdot e^{4x}+2}{1-C \cdot e^{4x}}$

Exercise 3-3:

$v + kyy' = 0 \quad k = \dfrac{c^2}{y_0^2}$

$\therefore vdv + k \cdot ydy = 0$

Appendix 1: Answers to Exercises

$$\therefore \frac{1}{2}v^2 + \frac{1}{2} \cdot k \cdot y^2 = \frac{1}{2}C$$

$$\therefore v^2 + \frac{c^2}{y_0^2}y^2 = C$$

$$\therefore y^2 = (C - v^2)\frac{y_0^2}{c^2}$$

$$\therefore y = \frac{y_0}{c}\sqrt{C - v^2}$$

When $v = 0$ $y = \frac{y_0}{c}\sqrt{C - 0} = y_0$

$$\therefore C = c^2$$

$$\therefore y = y_0\sqrt{1 - \frac{v^2}{c^2}}$$

Exercise 3-4:

$2xydx + x^2dy + 6xdx = \pi dy$

$\therefore (2xy + 6x)dx + (x^2 - \pi)dy = 0$

$\therefore 2x(y + 3)dx + (x^2 - \pi)dy = 0$

$\therefore \frac{2x}{x^2 - \pi}dx + \frac{1}{y+3}dy = 0$

$\therefore -\ln(\pi - x^2) - \ln(y + 3) = \ln C$

$\therefore \frac{1}{(y+3)(\pi - x^2)} = C$

$\therefore y = \frac{1}{C(\pi - x^2)} - 3$

Differential Equations in 24 Hours

Exercise 4-1:

$\frac{dy}{dx} + 5y = e^x \quad y = \frac{1}{6}e^x + C \cdot e^{-5x}$

$\therefore \frac{dy}{dx} = \frac{1}{6}e^x + C \cdot (-5)e^{-5x}$

$\therefore \frac{dy}{dx} + 5y = \frac{1}{6}e^x + C \cdot (-5)e^{-5x} + 5\left(\frac{1}{6}e^x + C \cdot e^{-5x}\right) = e^x$

$\therefore y = \frac{1}{6}e^x + C \cdot e^{-5x}$ is a solution to $\frac{dy}{dx} + 5y = e^x$.

Exercise 4-2:

$\frac{\partial^2 y}{\partial x^2} = \frac{\partial y}{\partial t} \qquad y(x,t) = e^{-t}\cos x$

$\therefore \frac{\partial y}{\partial x} = -e^{-t}\sin x \quad \frac{\partial y}{\partial t} = -e^{-t}\cos x$

$\therefore \frac{\partial^2 y}{\partial x^2} = -e^{-t}\cos x$

$\therefore \frac{\partial^2 y}{\partial x^2} = \frac{\partial y}{\partial t}$

$\therefore y(x,t) = e^{-t}\cos x$ is a solution of $\frac{\partial^2 y}{\partial x^2} = \frac{\partial y}{\partial t}$.

Exercise 4-3:

$y^2 + 1 = y \quad \text{first} - \text{order}$

$(y''')^5 + x^{11}(y^{(4)})^7 = e^{-13x} \quad \text{Fourth} - \text{order}$

$(x^7y^3)^{(6)} + (x^7y^3)^6 + y'''^{(4)} + e^{-10x}y^{(5)}x^{19} = y^{23}$ Seventh – order

Exercise 4-4:

$y'' + y = 2\cos x$ linear

$\dot{y} = y^2 + 1$ non – linear

$\dfrac{d^3y}{dx^3} - x^7 \dfrac{dy}{dx} = 0$ linear

$y' = x^2 + 1$ linear

Exercise 4-5:

$y' + \dfrac{1}{y+1} = 0 \quad y(1) = 0$

$\therefore (y+1)dy + dt = 0$

$\therefore \dfrac{1}{2}y^2 + y + t = C \therefore y^2 + 2y + 2t = C$

$\therefore (y+1)^2 + 2t - 1 = C$

$\therefore y + 1 = \pm\sqrt{-2t + 1 + C}$

$\therefore y = -1 \pm \sqrt{-2t + 1 + C}$

$\therefore y(1) = -1 \pm \sqrt{C - 1} = 0$

$\therefore \sqrt{C - 1} = 1$

$\therefore C = 2$

$\therefore y = -1 + \sqrt{-2t + 3}$

Since $-2t + 3 \geq 0$

$\therefore t \leq 1.5$

Exercise 5-1:

$8t + 2ty + (t^2 - 3)\dot{y} = 0$

$\therefore 8tdt + 2tydt + (t^2 - 3)dy = 0$

$\therefore \dfrac{2t}{(t^2-3)}dt + \dfrac{1}{4+y}dy = 0$

$\therefore \ln(t^2 - 3) + \ln(y + 4) = \ln C$

$\therefore \ln(y + 4) = \ln \dfrac{C}{(t^2-3)}$

$\therefore y + 4 = \dfrac{C}{(t^2-3)}$

$\therefore y = \dfrac{C}{(t^2-3)} - 4$

$\therefore y(2) = C - 4 = -3$

$\therefore C = 1$

Appendix 1: Answers to Exercises

$\therefore y = \frac{1}{(t^2-3)} - 4$

Exercise 5-2:

$y(0) = y_0 = 10$

$vdv + kydy = 0$

$\therefore \frac{1}{2}v^2 + k \cdot \frac{1}{2}y^2 = \frac{C}{2}$

$\therefore v^2 + k \cdot y^2 = C$

$\therefore 0 + ky_0^2 = C \quad c^2 + 0 = C$

$\therefore C = c^2 \quad k = \frac{c^2}{y_0^2}$

$\therefore v^2 + \frac{c^2}{y_0^2} \cdot y^2 = c^2$

$\therefore y = y_0 \sqrt{1 - \frac{v^2}{c^2}}$

$\therefore y(0.8c) = 10\sqrt{1 - \frac{0.64c^2}{c^2}} = 6m$

Exercise 6-1:

$y' - xy^{\frac{1}{2}} = 0 \quad y(0) = 0$

$$\therefore \frac{dy}{dx} = xy^{\frac{1}{2}} = f(x,y) \quad \text{continuous}$$

$$\frac{\partial f}{\partial y} = \frac{1}{2}xy^{-\frac{1}{2}} \quad \text{not continuous at } y = 0$$

∴ Picard's theorem does not guarantee existence and uniqueness.

Exercise 6-2:

$y' + xy = 0 \quad y(0) = 1$

$$\frac{dy}{dx} = -xy = f(x,y)$$

Continuous at (0,1) and any neighborhood in $\mathbf{R} \times \mathbf{R}$ containing (0,1)

$$\frac{\partial f}{\partial y} = -x$$

Continuous at (0,1) and any neighborhood in $\mathbf{R} \times \mathbf{R}$ containing (0,1)

∴ Picard's theorem guarantees uniquess.

Exercise 7-1:

$xdx + dy = 0$

$\therefore M = x \quad N = 1$

$$\therefore \frac{\partial M}{\partial y} = 0 \quad \frac{\partial N}{\partial x} = 0$$

Appendix 1: Answers to Exercises

$\therefore \dfrac{\partial f}{\partial x} = x \quad \dfrac{\partial f}{\partial y} = 1$

$\therefore f = \dfrac{1}{2}x^2 + \varphi_1(y) \quad f = y + \varphi_2(x)$

$\therefore f = \dfrac{1}{2}x^2 + y = C$

$\therefore y = -\dfrac{1}{2}x^2 + C$

Exercise 7-2:

$3x^2 y\, dx + (x^3 - y^2)dy = 0$

$\therefore M = 3x^2 y \quad N = x^3 - y^2$

$\therefore \dfrac{\partial M}{\partial y} = 3x^2 = \dfrac{\partial N}{\partial x} = 3x^2$

$\therefore \dfrac{\partial f}{\partial x} = 3x^2 y \quad \dfrac{\partial f}{\partial y} = x^3 - y^2$

$\therefore f = x^3 y + \varphi_1(y) \quad f = x^3 y - \dfrac{1}{3}y^3 + \varphi_2(x)$

$\therefore f = x^3 y - \dfrac{1}{3}y^3 = C$

Exercise 7-3:

$(2y + t\cos y)\dot{y} = -\sin y$

$\therefore \sin y\, dt + (2y + t\cos y)dy = 0$

$\therefore M = \sin y \quad N = 2y + t\cos y$

$\therefore \dfrac{\partial M}{\partial y} = \cos y = \dfrac{\partial N}{\partial t} = \cos y$

$\therefore \dfrac{\partial f}{\partial t} = \sin y \quad \dfrac{\partial f}{\partial y} = 2y + t\cos y$

211

Differential Equations in 24 Hours

$\therefore f = \sin y \, t + \varphi_1(y) \quad f = y^2 + t\sin y + \varphi_2(t)$

$\therefore f = y^2 + t\sin y = C$

Exercise 7-4:

$y \sec^2 x \, dx + \tan x \, dy = 0$

$\therefore M = y \sec^2 x \quad N = \tan x$

$\therefore \dfrac{\partial M}{\partial y} = \sec^2 x = \dfrac{\partial N}{\partial x} = \sec^2 x$

$\therefore \dfrac{\partial f}{\partial x} = y \sec^2 x \quad \dfrac{\partial f}{\partial y} = \tan x$

$\therefore f = y \tan x + \varphi_1(y) \quad f = y \tan x + \varphi_2(x)$

$\therefore f = y \tan x = C$

$\therefore C = y\left(\dfrac{\pi}{4}\right) \cdot \tan\left(\dfrac{\pi}{4}\right) = 1$

$\therefore y \tan x = 1$

$\therefore y = \cot x$

Exercise 7-5:

$y \ln y \, dx + x \, dy = 0$

$\therefore M = y \ln y \quad N = x$

$\therefore \dfrac{\partial M}{\partial y} = \ln y + 1 \neq \dfrac{\partial N}{\partial x} = 1$

$\therefore$ Not exact equation

$\therefore \dfrac{1}{x} dx + \dfrac{1}{y \ln y} dy = 0$

$\therefore \ln x + \ln(\ln y) = \ln C$

$\therefore \ln(\ln y) = \ln \dfrac{C}{x}$

$\therefore \ln y = \dfrac{C}{x}$

$\therefore y = e^{\frac{C}{x}}$

Exercise 8-1:

$\dfrac{dy}{dx} + 2y = e^{2x}$

$\therefore P(x) = 2$

$\therefore$ intergrating factor: $e^{\int^x 2 d\varphi} = e^{2x}$

$\therefore e^{2x}\dfrac{dy}{dx} + 2ye^{2x} = e^{4x}$

$\therefore \dfrac{d}{dx}(ye^{2x}) = e^{4x}$

$\therefore ye^{2x} = \dfrac{1}{4}e^{4x} + C$

$\therefore y = \dfrac{1}{4}e^{2x} + Ce^{-2x}$

Exercise 8-2:

$\cos^2 x \dfrac{dy}{dx} + y = e^{-\tan x}\cos^2 x$

$\therefore \dfrac{dy}{dx} + \dfrac{1}{\cos^2 x}y = e^{-\tan x}$

Differential Equations in 24 Hours

$\therefore P(x) = \dfrac{1}{\cos^2 x} = \sec^2 x$

$\therefore$ integrating factor: $e^{\int^x \sec^2 \varphi d\varphi} = e^{\tan x}$

$\therefore e^{\tan x} \dfrac{dy}{dx} + (\sec^2 x) e^{\tan x} y = 1$

$\therefore \dfrac{d}{dx}(e^{\tan x} y) = 1$

$\therefore e^{\tan x} y = x + C$

$\therefore y = \dfrac{x+C}{e^{\tan x}}$

Exercise 8-3:

$y dx + dy - (\sin x) dx = 0$

$\therefore y + \dfrac{dy}{dx} - \sin x = 0$

$\therefore \dfrac{dy}{dx} + y = \sin x$

$\therefore P(x) = 1$

$\therefore$ integrating factor: $e^{\int^x 1 d\varphi} = e^x$

$\therefore e^x \dfrac{dy}{dx} + e^x y = e^x \sin x$

$\therefore \dfrac{d}{dx}(e^x y) = e^x \sin x$

$\therefore e^x y = \dfrac{1}{2}(\sin x - \cos x) e^x + C$

$\therefore y = \dfrac{1}{2}(\sin x - \cos x) + C e^{-x}$

Appendix 1: Answers to Exercises

Exercise 8-4:

$$\frac{dy}{dx} + 2xy = e^{-x^2}$$

$$\therefore P(x) = 2x$$

$$\therefore \text{integrating factor: } e^{\int^x 2\varphi d\varphi} = e^{x^2}$$

$$\therefore e^{x^2}\frac{dy}{dx} + 2xe^{x^2}y = 1$$

$$\therefore \frac{d}{dx}\left(e^{x^2}y\right) = 1$$

$$\therefore e^{x^2}y = x + C$$

$$\therefore y = xe^{-x^2} + Ce^{-x^2}$$

Exercise 8-5:

$$L\frac{dj}{dt} + Rj = E$$

$$\therefore \frac{dj}{dt} + \frac{R}{L}j = \frac{E}{L}$$

$$\therefore P(t) = \frac{R}{L}$$

$$\therefore \text{integrating factor: } e^{\int^t \frac{R}{L}d\varphi} = e^{\frac{R}{L}t}$$

$$\therefore e^{\frac{R}{L}t}\frac{dj}{dt} + \frac{R}{L}j \cdot e^{\frac{R}{L}t} = e^{\frac{R}{L}t}\frac{E}{L}$$

$$\therefore \frac{d}{dt}\left(e^{\frac{R}{L}t} \cdot j\right) = e^{\frac{R}{L}t} \cdot \frac{E}{L}$$

$$\therefore e^{\frac{R}{L}t} \cdot j = \frac{E}{R}e^{\frac{R}{L}t} + C$$

Differential Equations in 24 Hours

$$\therefore j = \frac{E}{R} + \frac{C}{e^{\frac{R}{L}t}}$$

$$\therefore j(0) = \frac{E}{R} + C = 0$$

$$\therefore C = -\frac{E}{R}$$

$$\therefore j = \frac{E}{R} + \frac{-\frac{E}{R}}{e^{\frac{R}{L}t}} = \frac{E}{R}(1 - \frac{1}{e^{\frac{R}{L}t}})$$

Exercise 9-1:

$(x^2 + y^2)dx + 2xydy = 0$

$\therefore M(x,y) = (x^2 + y^2) \quad N(x,y) = 2xy$
$\therefore M(\tau x, \tau y) = (\tau x)^2 + (\tau y)^2 = \tau^2(x^2 + y^2) = \tau^2 M(x,y)$

$N(\tau x, \tau y) = 2(\tau x)(\tau y) = 2\tau^2 xy = \tau^2 N(x,y)$

$\therefore y = ux \therefore dy = udx + xdu$

$\therefore [x^2 + (ux)^2]dx + 2x(ux) \cdot (udx + xdu) = 0$

$\therefore x^2 dx + u^2 x^2 dx + 2u^2 x^2 dx + 2ux^3 du = 0$

$\therefore x^2 dx + 3u^2 x^2 dx + 2ux^3 du = 0$

$\therefore \frac{1}{x}dx + 3u^2 \cdot \frac{1}{x}dx + 2udu = 0$

$\therefore (1 + 3u^2) \cdot \frac{1}{x}dx + 2udu = 0$

$\therefore \frac{1}{x}dx + \frac{2u}{1+3u^2}du = 0$

Appendix 1: Answers to Exercises

$$\therefore \ln x + \frac{1}{3}\ln(1 + 3u^2) = \ln C$$

$$\therefore \ln(1 + 3u^2)^{\frac{1}{3}} = \ln\left(\frac{C}{x}\right)$$

$$\therefore (1 + 3u^2)^{\frac{1}{3}} = \frac{C}{x}$$

$$\therefore 1 + 3u^2 = \frac{C}{x^3}$$

$$\therefore u = \pm\sqrt{\frac{C}{3x^3} - \frac{1}{3}}$$

$$\therefore \frac{y}{x} = u = \pm\sqrt{\frac{C}{3x^3} - \frac{1}{3}}$$

$$\therefore y = \pm x \cdot \sqrt{\frac{C}{3x^3} - \frac{1}{3}} = \pm\sqrt{\frac{C}{x} - \frac{x^2}{3}}$$

Exercise 9-2:

$$\left(x\cos\frac{y}{x} - y\right)dx + x\,dy = 0$$

$$\therefore M(x,y) = x\cos\frac{y}{x} - y \quad N(x,y) = x$$

$$\therefore M(\tau x, \tau y) = \tau x \cos\frac{\tau y}{\tau x} - \tau y = \tau\left(x\cos\frac{y}{x} - y\right) = \tau M(x,y)$$

$$N(\tau x, \tau y) = \tau x = \tau N(x,y)$$

Let $y = ux$

$$\therefore dy = u\,dx + x\,du$$

$$\therefore \left(x\cos\frac{ux}{x} - ux\right)dx + x(u\,dx + x\,du) = 0$$

$\therefore x\cos u\, dx - u x\, dx + ux\, dx + x^2 du = 0$

$\therefore \dfrac{1}{x} dx + \dfrac{1}{\cos u} du = 0$

$\therefore \dfrac{1}{x} dx + \dfrac{\sec u(\sec u + \tan u)}{\sec u + \tan u} du = 0$

$\therefore \ln x + \ln(\sec u + \tan u) = \ln C$

$\therefore \sec u + \tan u = \dfrac{C}{x}$

$\therefore \dfrac{1}{\cos u} + \dfrac{\sin u}{\cos u} = \dfrac{C}{x}$

$\therefore \dfrac{1+\sin\frac{y}{x}}{\cos\frac{y}{x}} = \dfrac{C}{x}$

$\therefore x\left(1 + \sin\dfrac{y}{x}\right) = C \cdot \cos\dfrac{y}{x}$

Exercise 9-3:

$(x + 3y)dx + x\, dy = 0$

$\therefore M(x,y) = x + 3y \quad N(x,y) = x$

$\therefore M(\tau x, \tau y) = \tau x + 3\tau y = \tau M(x,y) \quad N(\tau x, \tau y) = \tau x = \tau N(x,y)$

$\therefore$ Let $y = ux$

$\therefore dy = u\, dx + x\, du$

$\therefore (x + 3ux)dx + x(u\, dx + x\, du) = 0$

$\therefore x\, dx + 4ux\, dx + x^2 du = 0$

$\therefore \dfrac{1}{x}(1 + 4u)dx + du = 0$

$$\therefore \frac{1}{x}dx + \frac{1}{1+4u}du = 0$$

$$\therefore \ln x + \frac{1}{4}\ln(1+4u) = \ln C$$

$$\therefore 1 + 4u = \frac{C}{x^4}$$

$$\therefore u = \frac{C}{4x^4} - \frac{1}{4}$$

$$\therefore \frac{y}{x} = u = \frac{C}{4x^4} - \frac{1}{4}$$

$$\therefore y = \frac{C}{4x^3} - \frac{x}{4} = C \cdot x^{-3} - \frac{x}{4}$$

Exercise 9-4:

$$x - \left(t + x\tan\frac{t}{x}\right)\dot{x} = 0$$

$$\therefore xdt - \left(t + x\tan\frac{t}{x}\right)dx = 0$$

$$\therefore M(t,x) = x \quad N(t,x) = t + x\tan\frac{t}{x}$$

$$\therefore M(\tau t, \tau x) = \tau x = \tau M(t,x) \quad N(\tau t, \tau x) = \tau t + \tau x \tan\frac{\tau t}{\tau x} = \tau N(t,x)$$

$$\therefore \text{Let } x = ut \therefore dx = udt + tdu$$

$$\therefore utdt - \left(t + ut \cdot \tan\frac{t}{ut}\right)(udt + tdu) = 0$$

$$\therefore u \cdot tdt - \left(u \cdot tdt + t^2 du + u^2 t \cdot \tan\frac{1}{u}dt + ut^2\tan\frac{1}{u}du\right) = 0$$

$$\therefore -t^2 du - u^2 t \cdot \tan\frac{1}{u}dt - ut^2\tan\frac{1}{u}du = 0$$

$$\therefore \frac{1}{u^2\tan\frac{1}{u}}du + \frac{1}{t}dt + \frac{1}{u}du = 0$$

$$\therefore -\ln\left(\sin\frac{1}{u}\right) + \ln t + \ln u = \ln C$$

$$\therefore \frac{u}{\sin\frac{1}{u}} = \frac{C}{t}$$

$$\therefore \frac{\frac{x}{t}}{\sin\frac{t}{x}} = \frac{C}{t}$$

$$\therefore \frac{x}{\sin\frac{t}{x}} = C$$

Exercise 10-1:

$$xy' + 2y = 4x^4 y^4$$

$$\therefore \frac{dy}{dx} + \frac{2}{x} y = 4x^3 y^4$$

$$\therefore n = 4 \quad (1-n)y^{-n} = -3y^{-4}$$

$$\therefore -3y^{-4}\frac{dy}{dx} + \left(-6\frac{1}{x}y^{-3}\right) = -12x^3$$

$$\therefore u = y^{1-n} = y^{-3}$$

$$\therefore du = -3y^{-4}\frac{dy}{dx}$$

$$\therefore du - 6\cdot\frac{1}{x}u = -12x^3$$

$\therefore$ integrating factor: $e^{\int^x -6\frac{1}{\varphi}d\varphi} = e^{-6\ln x} = x^{-6}$

$$\therefore x^{-6}du - 6\cdot x^{-7}u = -12x^{-3}$$

$$\therefore \frac{d}{dx}(x^{-6}\cdot u) = -12x^{-3}$$

Appendix 1: Answers to Exercises

$\therefore x^{-6} \cdot u = 6x^{-2} + C \therefore u = 6x^4 + Cx^6$

$\therefore y^{-3} = u = 6x^4 + Cx^6$

$\therefore y = x^{-1}(6x + C \cdot x^3)^{-\frac{1}{3}}$

Exercise 10-2:

$xy^3 y' + y^4 = \cos(x^4)$

$\therefore \frac{dy}{dx} + \frac{1}{x}y = \frac{\cos(x^4)}{x}y^{-3}$

$\therefore n = -3 \therefore (1-n)y^{-n} = 4y^3$

$\therefore 4y^3 \frac{dy}{dx} + \frac{4}{x}y^4 = \frac{4\cos(x^4)}{x}$

$\therefore u = y^{1-n} = y^4$

$\therefore du = 4y^3 \frac{dy}{dx}$

$\therefore du + \frac{4}{x}u = \frac{4\cos(x^4)}{x}$

$\therefore$ integrating factor: $e^{\int \frac{x4}{\varphi} d\varphi} = e^{4\ln x} = x^4$

$\therefore x^4 du + 4x^3 u = 4x^3 \cos(x^4)$

$\therefore \frac{d}{dx}(x^4 u) = 4x^3 \cos(x^4)$

$\therefore x^4 u = \sin(x^4) + C$

$\therefore u = \frac{\sin(x^4) + C}{x^4}$

$\therefore y^4 = u = \frac{\sin(x^4) + C}{x^4}$

$$\therefore y = \frac{1}{x}(\sin(x^4) + C)^{\frac{1}{4}}$$

Exercise 10-3:

$$\frac{dy}{dx} + \frac{1}{x}y = y^3 \cos\left(\frac{1}{x}\right)$$

$$\therefore n = 3 \therefore (1-n)y^{-n} = -2y^{-3}$$

$$\therefore -2y^{-3}\frac{dy}{dx} - \frac{2}{x}y^{-2} = -2\cos\left(\frac{1}{x}\right)$$

$$\therefore u = y^{1-n} = y^{-2}$$

$$\therefore du = -2y^{-3}\frac{dy}{dx}$$

$$\therefore du - \frac{2}{x}u = -2\cos\left(\frac{1}{x}\right)$$

$$\therefore \text{integrating factor: } e^{\int^x -\frac{2}{\varphi}d\varphi} = e^{-2\ln x} = x^{-2}$$

$$\therefore x^{-2}du - 2x^{-3}u = -2x^{-2}\cos\left(\frac{1}{x}\right)$$

$$\therefore \frac{d}{dx}(x^{-2}u) = -2x^{-2}\cos\left(\frac{1}{x}\right)$$

$$\therefore x^{-2}u = 2\sin\left(\frac{1}{x}\right) + C$$

$$\therefore u = 2x^2\sin\left(\frac{1}{x}\right) + Cx^2$$

$$\therefore y^{-2} = u = 2x^2\sin\left(\frac{1}{x}\right) + Cx^2$$

$$\therefore y = x^{-1}\left(2\sin\left(\frac{1}{x}\right) + C\right)^{-\frac{1}{2}}$$

Appendix 1: Answers to Exercises

Exercise 11-1:

$(3 + 4i)(3 - 4i) = 9 - 12i + 12i + 16 = 25 = \|3 + 4i\|^2$

$\|3 + 4i\| = 5$

Exercise 11-2:

$2\ln(i) = \ln(i)^2 = \ln(-1) = i\pi$

$\therefore \ln(i) = \frac{1}{2}i\pi$

$\therefore \frac{\ln(i)}{i} = \frac{1}{2}\pi$

$\therefore \cos\left(\frac{\ln(i)}{i}\right) = \cos\left(\frac{1}{2}\pi\right) = 0$

Exercise 11-3:

$1 = e^{i2\pi n}$ where $n = \cdots -2, -1, 0, 1, 2 \cdots$

$\therefore 1^{\frac{1}{4}} = e^{i\frac{\pi}{2}n}$ where $n = \cdots -2, -1, 0, 1, 2 \cdots$

$\therefore 1^{\frac{1}{4}} = \{i, -1, -i, 1\}$

Exercise 11-4:

$h(t)^* h(t) = e^{i\frac{2\pi}{1024}t} e^{-i\frac{2\pi}{1024}t} = e^0 = 1$

Exercise 11-5:

$\|h(t)\|^2 = \int_{-\pi}^{\pi} h(t)h*(t)\,dt = \int_{-\pi}^{\pi} e^{i\frac{2\pi}{1024}t} e^{-i\frac{2\pi}{1024}t}\,dt = \int_{-\pi}^{\pi} 1\,dt = 2\pi$

$\|h(t)\| = \sqrt{2\pi}$

Exercise 11-6: $FT(e^{j\omega t}) = 2\pi$

$\|h(t)\|^2 = \int_{-\pi}^{\pi} h(t)h^*(t)\,dt = \int_{-\pi}^{\pi} e^{i\frac{2\pi}{1024}t} e^{-i\frac{2\pi}{1024}t}\,dt = \int_{-\pi}^{\pi} 1\,dt = 2\pi$

$FT(e^{j\alpha t}) = \int_{-\pi}^{\pi} e^{j\alpha t} e^{-j\alpha t}\,dt = \int_{-\pi}^{\pi} 1\,dt = 2\pi$

Exercise 11-7:
$i^i = e^{-(\pi/2 + 2\pi)} = .000388$
$i^i = e^{-(\pi/2 + 4\pi)} = 7.25 \times 10^{-7}$

Exercise 11–8: $5 = (1 - 2i)(1 + 2i)$

5 is prime; however, it is not a **Gaussian prime**.

Exercise 12-1:

$\ddot{y} + \omega^2 y = 0$ has solutions

$\left\{ \sin(\omega t), \cos(\omega t), \sqrt{2}\cos\left(\omega t - \frac{\pi}{4}\right) \right\}$

$\sqrt{2}\cos\left(\omega t - \frac{\pi}{4}\right) = \sqrt{2}\left[\cos(\omega t)\cos\left(\frac{\pi}{4}\right) + \sin(\omega t)\sin\left(\frac{\pi}{4}\right)\right] =$

$\cos(\omega t) + \sin(\omega t)$

Exercise 12-2:

Appendix 1: Answers to Exercises

$$W = \begin{vmatrix} y_1 & y_2 & y_3 \\ y_1' & y_2' & y_3' \\ y_1'' & y_2'' & y_3'' \end{vmatrix} = 0$$

because the three solutions are not independent.

Exercise 12-3:

$$\frac{d^2j}{dt^2} + \frac{1}{LC}j = \frac{w_0}{L}\cos(wt)$$

$$\therefore \frac{d^2j}{dt^2} + \frac{1}{LC}j_h = 0$$

$$\therefore \frac{dj}{dt} = -C_1 \sin\left(\frac{1}{\sqrt{LC}}t\right) \cdot \frac{1}{\sqrt{LC}} + C_2 \cos\left(\frac{1}{\sqrt{LC}}t\right) \cdot \frac{1}{\sqrt{LC}}$$

$$\therefore \frac{d^2j}{dt^2} = -\frac{1}{LC}\left(C_1 \cos\left(\frac{1}{\sqrt{LC}}t\right) + C_2 \sin\left(\frac{1}{\sqrt{LC}}t\right)\right) = -\frac{1}{LC}j_h$$

$$\therefore \frac{d^2j}{dt^2} + \frac{1}{LC}j = -\frac{1}{LC}j_h + \frac{1}{LC}j_h = 0$$

Exercise 12-4:

$$j_p = \frac{\omega_0 C}{1 - \omega_0^2 LC}\cos(\omega_0 t)$$

$$\therefore \frac{dj_p}{dt} = \frac{\omega_0 C}{1-\omega_0^2 LC}(-\sin(\omega_0 t)) \cdot \omega_0 = -\frac{\omega_0^2 C}{1-\omega_0^2 LC}\sin(\omega_0 t)$$

$$\therefore \frac{d^2 j_p}{dt^2} = -\frac{\omega_0^2 C}{1-\omega_0^2 LC} \cdot \cos(\omega_0 t) \cdot \omega_0 = -\frac{\omega_0^3 C}{1-\omega_0^2 LC}\cos(\omega_0 t)$$

$$\therefore -\frac{\omega_0^3 C}{1-\omega_0^2 LC}\cos(\omega_0 t) + \frac{1}{LC}\frac{\omega_0 C}{1-\omega_0^2 LC}\cos(\omega_0 t) =$$

$$\left[-\omega_0^2 + \frac{1}{LC}\right]\frac{\omega_0 C}{1 - \omega_0^2 LC}\cos(\omega_0 t)$$

Differential Equations in 24 Hours

$$= \frac{\omega_0}{L} \cos(\omega_0 t)$$

Exercise 12-5:

$$j_g = j_h + j_p = C_1 \cos\left(\frac{1}{\sqrt{LC}}t\right) + C_2 \sin\left(\frac{1}{\sqrt{LC}}t\right)$$

$$+ \frac{\omega_0 C}{1 - \omega_0^2 LC} \cos(\omega_0 t)$$

Exercise 12-6:

The amplitude of $j_p = \dfrac{1}{\sqrt{R^2 + \left[\dfrac{1}{\omega_0 C} - \omega_0 L\right]^2}} \cos(\omega_0 t)$ is

maximized when we choose C to make the denominator as small as possible. We do this by setting:

$$\left[\frac{1}{\omega_0 C} - \omega_0 L\right] = 0$$

Solving for C, we have:

$$C_{resonance} = \frac{1}{\omega_0^2 L}$$

Exercise 12-7:

$$y(L) = C_2 \sin(\omega_m L) = 0,$$

which requires that $\omega_m L = q\pi$, $q = ..., -2, -1, 0, 1, 2, ...$

The lowest non-zero frequency occurs when q = 1.

Thus: $\omega_1 = \dfrac{\pi}{L}$

Exercise 12-8:

$$y(\frac{L}{2}) = C_2 \sin(\omega_m \frac{L}{2}) = 0$$

which requires that $\omega_m \frac{L}{2} = q\pi, \quad q = ...,-2,-1,0,1,2,...$

The lowest frequency occurs for q = 1:

Which gives: $\omega_1 = \frac{2\pi}{L}$

Exercise 13-1:

$$\frac{d^2y}{dx^2} - 3\frac{dy}{dx} + 2y = 0$$

Let $y = e^{mx}$

$m^2 e^{mx} - 3m e^{mx} + 2 e^{mx} = 0$

$m^2 - 3m + 2 = 0 \quad (m-1)(m-2) = 0$

$\therefore y_1 = e^x \quad y_2 = e^{2x}$

$\therefore y = C_1 e^x + C_2 e^{2x}$

Exercise 13-2:

$\ddot{y} - 2\dot{y} + 5y = 0$

Let $y = e^{mx}$

$m^2 e^{mx} - 2m\, e^{mx} + 5 e^{mx} = 0$

$m^2 - 2m + 5 = 0$

$(m - 1)^2 + 4 = 0$

$m = 1 \pm 2i$

$\therefore y = e^x(C_1 \cos(2x)\, C_2 \sin(2x))$

Exercise 13-3:

$y'' - 8y' + 16y = 0$

Let $y = e^{mx}$

$m^2 e^{mx} - 8m e^{mx} + 16 e^{mx} = 0$

$m^2 - 8m + 16 = 0$

$\therefore m = 4$

$\therefore y_1 = e^{4x} \qquad y_2 = u_{(x)} e^{4x}$

$\therefore y_2' = u' e^{4x} + 4u e^{4x}$

$\therefore y_2'' = u'' e^{4x} + 4u' e^{4x} + 4u' e^{4x} + 16u e^{4x}$

$\therefore u'' e^{4x} + 4u' e^{4x} + 4u' e^{4x} + 16u e^{4x} - 8u' e^{4x} - 32u e^{4x} + 16u e^{4x} = 0$

$\therefore u'' e^{4x} = 0$

$\therefore u'' = 0$

$\therefore u_{(x)} = C_1 x + C_2$

$\therefore y_2 = (C_1 x + C_2) e^{4x}$

$\therefore y = C_1 x e^{4x} + C_2 e^{4x}$

Appendix 1: Answers to Exercises

Exercise 13-4:

$y'' - (2 + \pi)y' + 2\pi y = 0$

Let $y = e^{mx}$

$m^2 e^{mx} - (2 + \pi)m e^{mx} + 2\pi e^{mx} = 0$

$\therefore m^2 - (2 + \pi)m + 2\pi = 0$

$\therefore \left[m - \left(1 + \frac{\pi}{2}\right)\right]^2 = \left(1 - \frac{\pi}{2}\right)^2$

$\therefore m - \left(1 + \frac{\pi}{2}\right) = \pm\left(1 - \frac{\pi}{2}\right)$

$\therefore m_1 = 2 \quad m_2 = \pi$

$\therefore y_1 = e^{2x} \quad y_2 = e^{\pi x}$

$\therefore y = C_1 e^{2x} + C_2 e^{\pi x}$

$\therefore y_{(0)} = 1 = C_1 + C_2 \quad y_{(0)}' = 2C_1 + \pi C_2 = \pi$

$\therefore C_1 = 0 \quad C_2 = 1$

$\therefore y = e^{\pi x}$

Exercise 14-1:

$y'' + y'y = 0$

Let $y' = u$ and $y'' = u\dfrac{du}{dy}$

$\therefore u\dfrac{du}{dy} + u \cdot y = 0$

Differential Equations in 24 Hours

$\therefore du + y\,dy = 0$

$\therefore y' + \frac{1}{2}y^2 = C_0$

$\therefore \frac{dy}{dx} = -\frac{1}{2}y^2 + C_0$

$\therefore \frac{dy}{dx} = -\frac{1}{2}(y + \sqrt{2C_0})(y - \sqrt{2C_0})$

$\therefore dy = -\frac{1}{2}(y + C_1)(y - C_2)dx$

$\therefore \frac{dy}{(y+C_1)(y-C_1)} = -\frac{1}{2}dx$

Let $\frac{1}{(y+C_1)(y-C_1)} = \frac{A}{y+C_1} + \frac{B}{y-C_1}$

$\therefore A(y - C_1) + B(y + C_1) = 1$

$\therefore (A + B)y + (A - B)C_1 = 1$

$\therefore \begin{cases} A + B = 0 \\ A - B = \frac{1}{C_1} \end{cases} \therefore \begin{cases} A = \frac{1}{2C_1} \\ B = -\frac{1}{2C_1} \end{cases}$

$\therefore -\frac{\frac{1}{2C_1}}{y+C_1}dy + \frac{\frac{1}{2C_1}}{y-C_1}dy = \frac{1}{2}dx$

$\therefore -\ln(y + C_1)^{\frac{1}{2C_1}} + \ln(y - C_1)^{\frac{1}{2C_1}} = \frac{1}{2}x + C$

$\therefore \left(\frac{y-C_1}{y+C_1}\right)^{\frac{1}{2C_1}} = e^{\frac{1}{2}x+C}$

$\therefore \frac{y-C_1}{y+C_1} = e^{C_1 x + 2CC_1} = e^{C_2 + C_1 x}$

If $\frac{y-C_1}{y+C_1} = Q$

230

Appendix 1: Answers to Exercises

then $y - C_1 = Qy + QC_1$

$\therefore (1-Q)y = (1+Q)C_1$

$\therefore y = \dfrac{1+Q}{1-Q} C_1$

$\therefore y = C_1 \dfrac{1+e^{C_2+C_1 x}}{1-e^{C_2+C_1 x}}$

Exercise 14-2:

$y'' - y' + \dfrac{1}{4} y = 0$

Let $y = e^{mx}$

$\therefore m^2 e^{mx} - m e^{mx} + \dfrac{1}{4} e^{mx} = 0$

$\therefore m^2 - m + \dfrac{1}{4} = 0$

$\therefore 4m^2 - 4m + 1 = 0$

$\therefore (2m-1)^2 = 0$

$\therefore m = \dfrac{1}{2}$

$\therefore y_1 = e^{\frac{1}{2}x} \quad y_2 = u_{(x)} e^{\frac{1}{2}x}$

$\therefore y_2' = u' e^{\frac{1}{2}x} + \dfrac{1}{2} u e^{\frac{1}{2}x}$

$\therefore y_2'' = u'' e^{\frac{1}{2}x} + \dfrac{1}{2} u' e^{\frac{1}{2}x} + \dfrac{1}{2} u' e^{\frac{1}{2}x} + \dfrac{1}{4} u e^{\frac{1}{2}x}$

$\therefore u'' e^{\frac{1}{2}x} + \dfrac{1}{2} u' e^{\frac{1}{2}x} + \dfrac{1}{2} u' e^{\frac{1}{2}x} + \dfrac{1}{4} u e^{\frac{1}{2}x} - u' e^{\frac{1}{2}x} - \dfrac{1}{2} u e^{\frac{1}{2}x} + \dfrac{1}{4} u e^{\frac{1}{2}x} = 0$

$\therefore u'' e^{\frac{1}{2}x} = 0$

231

Differential Equations in 24 Hours

$\therefore u'' = 0$

$\therefore u = C_1 x + C_2$

$\therefore y_2 = (C_1 x + C_2) e^{\frac{1}{2}x}$

$\therefore y_g = C_1 x e^{\frac{1}{2}x} + C_2 e^{\frac{1}{2}x}$

Exercise 14-3

$xy'' + 2y' = x^2$

Let $y = kx^3 \therefore y' = 3kx^2$

$y'' = 6kx$

$\therefore x \cdot (6kx) + 2(3kx^2) = x^2$

$\therefore 6kx^2 + 6kx^2 = x^2$

$\therefore 12k = 1$

$\therefore k = \frac{1}{12}$

$\therefore y_p = \frac{1}{12} x^3$

Exercise 14-4:

Look for a second solution in the form $y_2 = u\, y_1 = u\, x$

$y_2' = u'x + u$

$y_2'' = u''x + u' + u' = u''x + 2u'$

Plugging in, we have:

$$x^2(u''x + 2u') - x(u'x + u) + ux = 0$$

$$x^3u'' + 2x^2u' - u'x^2 - xu + ux = 0$$

$$x^3u'' + 2x^2u' - u'x^2 = 0$$

$$xu'' + u' = 0$$

Let: $u' = w,\ u'' = w'$

$$xw' + w = 0$$

$$\frac{dw}{w} + \frac{dx}{x} = 0$$

$$\ln(wx) = \ln C$$

$$wx = C$$

$$u' = \frac{C}{x} \Rightarrow u = C_1 \ln x + C_2$$

Or simply $u = \ln x$

$$y_2 = uy_1 = x \ln x$$

Exercise 15-1:

$$y'' + 3y' - 10y = 6e^{4x}$$

The first step is to solve the homogeneous equation: $y'' + 3y' - 10y = 0$

Letting $y = e^{mx}$, we find the characteristic equation:

Differential Equations in 24 Hours

$m^2 + 3m - 10 = 0 \Rightarrow (m-2)(m+5) = 0$

This has two solutions: $S_h = \{e^{2x}, e^{-5x}\}$

Step 2.) The terms of g(x) with all its derivatives consists of the single item e^{4x}.

Step 3.) $y_{p\_temp} = Ce^{4x}$

Step 4.) Comparing the terms of y_{p_temp} to the elements of S_h, we find nothing in common; therefore, we do not need to multiply by a power of x. Thus:

$y_p = Ce^{4x}$

Step 5.) Plugging in we have:

$16Ce^{4x} + 12Ce^{4x} - 10Ce^{4x} = 6e^{4x}$

Thus C = 1/3.

Thus we have: $y_g = C_1 e^{-5x} + C_2 e^{2x} + \frac{1}{3} e^{4x}$

Exercise 15-2:

$y'' + 4y = 3\sin x$

Step 1.) $S_h = \{\sin 2x, \cos 2x\}$

Step 2.) sinx, cosx

Step 3.) $y_{p\_temp} = A\sin x + B\cos x$

Step 4.) $y_p = A\sin x + B\cos x$

Step 5.) $y_p' = A\cos x - B\sin x$, $y_p'' = -A\sin x - B\cos x$

$-A\sin x - B\cos x + 4A\sin x + 4B\cos x = 3\sin x$

234

$3A\sin x + 3B\cos x = 3\sin x \Rightarrow A = 1,\ B = 0$

$y_p = \sin x$

$y_g = C_1 \cos 2x + C_2 \sin 2x + \sin x$

Exercise 15-3:

$2y'' + 3y' + y = \cos x$

Step 1.) $2m^2 + 3m + 1 = 0$

$(2m + 1)(m + 1) = 0 \Rightarrow m = -\tfrac{1}{2}, -1$

$S_h = \left\{ e^{-\tfrac{1}{2}x},\ e^{-x} \right\}$

Step 2.) cosx, sinx

Step 3.) $y_{p\_temp} = A\cos x + B\sin x$

Step 4.) $y_p = A\cos x + B\sin x$

Step 5.) $y'_p = -A\sin x + B\cos x \Rightarrow y''_p = -A\cos x - B\sin x$

$-2A\cos x - 2B\sin x - 3A\sin x + 3B\cos x + A\cos x + B\sin x = \cos x$

$(3B - A)\cos x - (B + 3A)\sin x = \cos x$

$\left. \begin{array}{l} -A + 3B = 1 \\ 3A + B = 0 \end{array} \right\} \Rightarrow \begin{bmatrix} -1 & 3 & 1 \\ 3 & 1 & 0 \end{bmatrix} \Rightarrow \begin{bmatrix} 1 & -3 & -1 \\ 0 & 10 & 3 \end{bmatrix}$

$\begin{bmatrix} 1 & -3 & -1 \\ 0 & 1 & 3/10 \end{bmatrix} \Rightarrow \begin{bmatrix} 1 & 0 & -1/10 \\ 0 & 1 & 3/10 \end{bmatrix}$

Differential Equations in 24 Hours

$$y_p = -\tfrac{1}{10}\cos x + \tfrac{3}{10}\sin x$$

$$y_g = C_1 e^{-\tfrac{1}{2}x} + C_2 e^{-x} - \tfrac{1}{10}\cos x + \tfrac{3}{10}\sin x$$

Exercise 15-4:

$$y'' - 2y' - 3y = 2e^{3x} + 8xe^{3x}$$

Step 1.) $m^2 - 2m - 3 = 0 \Rightarrow (m+1)(m-3) = 0 \Rightarrow m = -1, 3$

$$S_h = \{e^{-x}, e^{3x}\}$$

Step 2.) e^{3x}, xe^{3x}

Step 3.) $y_{p\_temp} = Ae^{3x} + Bxe^{3x}$

Step 4.) $y_p = Axe^{3x} + Bx^2 e^{3x}$

Step 5.) $y'_p = Ae^{3x} + 3Axe^{3x} + 2Bxe^{3x} + 3Bx^2 e^{3x}$

$$y''_p = 3Ae^{3x} + 3Ae^{3x} + 9Axe^{3x} + 2Be^{3x} + 6Bxe^{3x} + 6Bxe^{3x} + 9Bx^2 e^{3x}$$

$$y''_p = (6A + 2B)e^{3x} + (12B + 9A)xe^{3x} + 9Bx^2 e^{3x}$$

$(6A + 2B) + (12B + 9A)x + 9Bx^2 - 2A - 6Ax - 4Bx - 6Bx^2 - 3Ax - 3Bx^2$
$= 2 + 8x$
$4A + 2B + (6A + 8B)x = 2 + 8x$

$4A + 2B = 2, \quad 8B + 6A = 8$

$$\begin{bmatrix} 2 & 1 & 1 \\ 3 & 4 & 4 \end{bmatrix} \Rightarrow \begin{bmatrix} 2 & 1 & 1 \\ 1 & 3 & 3 \end{bmatrix} \Rightarrow \begin{bmatrix} 1 & 3 & 3 \\ 2 & 1 & 1 \end{bmatrix} \Rightarrow$$

Appendix 1: Answers to Exercises

$$\begin{bmatrix} 1 & 3 & 3 \\ 0 & -5 & -5 \end{bmatrix} \Rightarrow \begin{bmatrix} 1 & 3 & 3 \\ 0 & 1 & 1 \end{bmatrix} \Rightarrow \begin{bmatrix} 1 & 0 & 0 \\ 0 & 1 & 1 \end{bmatrix}$$

$A = 0, B = 1$

$y_p = x^2 e^{3x}$

$y_g = C_1 e^{3x} + C_2 e^{-x} + x^2 e^{3x}$

Exercise 16-1:

$y'' + 9y = \tan 3x$

$\therefore \{y1, y2\} = \{\cos 3x, \sin 3x\}$

$\therefore W = \begin{vmatrix} \cos 3x & \sin 3x \\ -3\sin 3x & 3\cos 3x \end{vmatrix} = 3(\cos 3x)^2 + 3(\sin 3x)^2 = 3$

$\therefore u_1' = \dfrac{-\tan 3x \cdot \sin 3x}{3} = -\dfrac{1}{3} \dfrac{(\sin 3x)^2}{\cos 3x} = -\dfrac{1}{3} \cdot \dfrac{1 - \cos 3x^2}{\cos 3x} = -\dfrac{1}{3}\sec 3x + \dfrac{1}{3}\cos 3x$

$u_2' = \dfrac{\tan 3x \cdot \cos 3x}{3} = \dfrac{1}{3}\sin 3x$

$\therefore u_1 = -\dfrac{1}{3}\int \sec 3x\, dx + \dfrac{1}{3}\int \cos 3x\, dx = -\dfrac{1}{9}\ln|\sec 3x + \tan 3x| + \dfrac{1}{9}\sin 3x$

$u_2 = \int \dfrac{1}{3}\sin 3x\, dx = -\dfrac{1}{9}\cos 3x$

$\therefore y_p = \cos 3x \left(-\dfrac{1}{9}\ln|\sec 3x + \tan 3x| + \dfrac{1}{9}\sin 3x\right) + \sin 3x \left(-\dfrac{1}{9}\cos 3x\right)$

Differential Equations in 24 Hours

$$= -\frac{1}{9}\ln|\sec 3x + \tan 3x| \cdot \cos 3x$$

Exercise 16-2:

$y'' - y = e^{2x}$

Let $y = e^{mx}$

$\therefore m^2 - 1 = 0$

$\therefore m = \pm 1$

$\therefore y_1 = e^x \quad y_2 = e^{-x}$

$\therefore W = \begin{vmatrix} e^x & e^{-x} \\ e^x & -e^{-x} \end{vmatrix} = -1 - 1 = -2$

$\therefore u_1' = \dfrac{-e^{2x} \cdot e^{-x}}{-2} = \dfrac{1}{2}e^x \quad u_2' = \dfrac{e^{2x} \cdot e^x}{-2} = -\dfrac{1}{2}e^{3x}$

$\therefore u_1 = \dfrac{1}{2}e^x \quad u_2 = -\dfrac{1}{6}e^{3x}$

$\therefore y_p = e^x \cdot \dfrac{1}{2}e^x + e^{-x} \cdot \left(-\dfrac{1}{6}e^{3x}\right) = \dfrac{1}{3}e^{2x}$

$\therefore y_g = C_1 e^x + C_2 e^{-x} + \dfrac{1}{3}e^{2x}$

Exercise 16-3:

$y'' + 2y' + y = e^{3x}$

$\therefore m^2 + 2m + 1 = 0$

$\therefore (m+1)^2 = 0$

$\therefore m = -1$

$\therefore y_1 = e^{-x} \quad y_2 = xe^{-x}$

Appendix 1: Answers to Exercises

$$\therefore W = \begin{vmatrix} e^{-x} & xe^{-x} \\ -e^{-x} & (e^{-x} - xe^{-x}) \end{vmatrix} = e^{-2x}$$

$$\therefore u_1' = \frac{-e^{3x} \cdot xe^{-x}}{e^{-2x}} = -xe^{4x} \quad u_2' = \frac{e^{3x} \cdot e^{-x}}{e^{-2x}} = e^{4x}$$

$$\therefore u_1 = \int -xe^{4x} dx = -\frac{1}{4}xe^{4x} + \frac{1}{16}e^{4x} \quad u_2 = \int e^{4x} dx = \frac{1}{4}e^{4x}$$

$$\therefore y_p = e^{-x}\left(-\frac{1}{4}xe^{4x} + \frac{1}{16}e^{4x}\right) + xe^{-x} \cdot \frac{1}{4}e^{4x} = \frac{1}{16}e^{3x}$$

$$\therefore y_g = C_1 e^{-x} + C_2 xe^{-x} + \frac{1}{16}e^{3x}$$

Exercise 16-4:

$$y'' + y = \csc x$$

$$\therefore y_1 = \cos x \quad y_2 = \sin x$$

$$\therefore W = \begin{vmatrix} \cos x & \sin x \\ -\sin x & \cos x \end{vmatrix} = 1$$

$$\therefore u_1' = \frac{-\csc x \cdot \sin x}{1} = -1 \quad u_2' = \frac{\csc x \cdot \cos x}{1} = \frac{1}{\sin x}\cos x$$

$$\therefore u_1 = \int -1 \, dx = -x \quad u_2 = \int \frac{1}{\sin x}\cos x \, dx = \ln(\sin x)$$

$$\therefore y_p = -x \cdot \cos x + \sin x \cdot \ln(\sin x)$$

$$\therefore y_g = C_1 \cos x + C_2 \sin x - x \cdot \cos x + \sin x \cdot \ln(\sin x)$$

Exercise 17-1:

$$x^2 y'' + 13xy' + 20y = 0$$

let $y = x^m$

239

$\therefore m(m-1) + 13m + 20 = 0$

$\therefore m^2 + 12m + 20 = 0$

$\therefore (m+10)(m+2) = 0$

$\therefore y_1 = x^{-10} \quad y_2 = x^{-2}$

$\therefore y_h = C_1 x^{-10} + C_2 x^{-2} = y_g$

Exercise 17-2:

$x^2 y'' + 3xy' + y = \pi$

Let $y = x^m$

$\therefore m(m-1) + 3m + 1 = 0$

$\therefore m^2 + 2m + 1 = 0$

$\therefore (m+1)^2 = 0 \quad \therefore$ double root

$\therefore y_h = C_1 x^{-1} + C_2 x^{-1} \ln x$

Since $x^2 y'' + 3xy' + y = \pi$

$\therefore y_g = y_h + \pi = C_1 x^{-1} + C_2 x^{-1} \ln x + \pi$

Exercise 17-3:

$4x^2 y'' + 8xy' + 5y = 0$

Let $y = x^m$

$\therefore 4m(m-1) + 8m + 5 = 0$

$\therefore 4m^2 + 4m + 5 = 0$

Appendix 1: Answers to Exercises

$\therefore (2m+1)^2 = -4$

$\therefore m = -\frac{1}{2} \pm i$

$\therefore y_h = x^{-\frac{1}{2}}[C_1 \cos(\ln x) + C_2 \sin(\ln x)] = y_g$

Exercise 18-1:

$y'' + 49y = 0$

Let $y = \sum_{n=0}^{\infty} C_n x^n$

$\therefore y' = \sum_{n=0}^{\infty} n \cdot C_n x^{n-1}$

$\therefore y'' = \sum_{n=0}^{\infty} n(n-1) C_n x^{n-2}$

$\therefore \sum_{n=0}^{\infty} n(n-1) C_n x^{n-2} + 49 \sum_{n=0}^{\infty} C_n x^n = 0$

$\therefore (0 + 0 + \sum_{n=0}^{\infty} (n+2)(n+1) C_{n+2} x^n) + \sum_{n=0}^{\infty} 49 C_n x^n = 0$

$\therefore \sum_{n=0}^{\infty} [(n+1)(n+2) C_{n+2} + 49 C_n] x^n = 0$

$\therefore (n+1)(n+2) C_{n+2} + 49 C_n = 0$

$\therefore C_{n+2} = \frac{-49}{(n+1)(n+2)} C_n = \frac{(7i)^2}{(n+1)(n+2)} C_n$

$\therefore$ Even terms: $C_2 = \frac{(7i)^2}{1 \times 2} C_0$

$C_4 = \frac{(7i)^2}{3 \times 4} C_2 = \frac{(7i)^4}{1 \times 2 \times 3 \times 4} C_0$

$C_6 = \frac{(7i)^6}{6!} C_0$

$C_{2k} = \frac{(7i)^{2k}}{(2k)!} C_0$

Differential Equations in 24 Hours

Odd terms: $\quad C_3 = \dfrac{(7i)^2}{2\times 3} C_1 = \dfrac{(7i)^3}{2\times 3}\cdot\dfrac{C_1}{7i}$

$C_5 = \dfrac{(7i)^2}{4\times 5} C_3 = \dfrac{(7i)^5}{2\times 3\times 4\times 5}\dfrac{C_1}{7i}$

$C_7 = \dfrac{(7i)^7}{7!}\dfrac{C_1}{7i}$

$\vdots$

$C_{2k+1} = \dfrac{(7i)^{2k+1}}{(2k+1)!}\dfrac{C_1}{7i}$

$\therefore y = \sum_{n=0}^{\infty} C_n x^n$

$= \sum_{n=0}^{\infty} x^n \dfrac{(7i)^n}{(n)!} C_0 + \sum_{n=0}^{\infty} x^n \dfrac{(7i)^n}{(n)!}\dfrac{C_1}{7i}$

$= C_0 \sum_{n=0}^{\infty} \dfrac{1}{n!}(7ix)^n + \dfrac{C_1}{7i}\sum_{n=0}^{\infty} \dfrac{1}{n!}(7ix)^n$

$= C_0 e^{7ix} + \dfrac{C_1}{7i} e^{7ix}$

$= C_1 \cos 7x + C_2 \sin 7x$

Exercises 18-2, 3, 4:

$y'' - 2xy' + 2py = 0$

Let $y = \sum_{n=0}^{\infty} C_n x^n$

$\therefore y' = \sum_{n=0}^{\infty} n\cdot C_n x^{n-1}$

$\therefore y'' = \sum_{n=0}^{\infty} n(n-1) C_n x^{n-2}$

$\therefore \sum_{n=0}^{\infty} n(n-1) C_n x^{n-2} - \sum_{n=0}^{\infty} 2n\cdot C_n x^n + \sum_{n=0}^{\infty} 2p C_n x^n = 0$

$\therefore \sum_{n=0}^{\infty}(n+2)(n+1) \cdot C_{n+2}x^n - \sum_{n=0}^{\infty} 2n \cdot C_n x^n + \sum_{n=0}^{\infty} 2pC_n x^n = 0$

$\therefore \sum_{n=0}^{\infty}[(n+2)(n+1) \cdot C_{n+2} + (2p - 2n) \cdot C_n]x^n = 0$

$\therefore C_{n+2} = \dfrac{2(n-p)}{(n+1)(n+2)} C_n$

Even terms: $C_2 = -pC_0$

$C_4 = \dfrac{2(2-p)}{4\cdot 3} C_2 = \dfrac{2^2 p(p-2)}{4!} C_0$

$C_6 = -\dfrac{2^3 p(p-2)(p-4)}{6!} C_0 \quad \cdots$

Odd terms: $C_3 = -\dfrac{2(p-1)}{3!} C_1$

$C_5 = \dfrac{2(3-p)}{5\cdot 4} C_3 = \dfrac{2^2 (p-1)(p-3)}{5!} C_1$

$C_7 = -\dfrac{2^3 (p-1)(p-3)(p-5)}{7!} C_1 \quad \cdots$

$y = C_0 y_1 + C_1 y_2$ where

$y_1 = 1 - \dfrac{2p}{2!}x^2 + \dfrac{2^2 p(p-2)}{4!}x^4 - \dfrac{2^3 p(p-2)(p-4)}{6!}x^6 + \cdots$

and $y_2 = x - \dfrac{2(p-1)}{3!}x^3 + \dfrac{2^2(p-1)(p-3)}{5!}x^5 - \dfrac{2^3(p-1)(p-3)(p-5)}{7!}x^7 + \cdots$

$\therefore$ when $p = 0$ $y_1 = 1$

when $p = 1$ $y_2 = x$

when $p = 2$ $y_1 = 1 - 2x^2$

Exercise 20-1:

$y(t) = 2\sin t\cos t = \sin(2t)$

$\therefore Y(s) = \dfrac{2}{4+s^2}$

Exercise 20-2:

$y(t) = t^6(\sin 7t)^2 + t^6(\cos 7t)^2 = t^6$

$\therefore Y(s) = \dfrac{6!}{s^7} = \dfrac{720}{s^7}$

Exercise 20-3:

$Y(s) = \dfrac{1}{s} - \dfrac{1}{s^2+1}$

$\therefore y(t) = 1 - \sin t$

Exercise 20-4:

$2\dfrac{dy}{dt} + y = 0 \quad y(0) = -3$

$\therefore 2sY(s) - 2y(0) + Y(s) = 0$

$\therefore (2s+1)Y(s) = 2y(0) = -6$

$\therefore Y(s) = \dfrac{-6}{2s+1} = -3 \cdot \dfrac{1}{s - \tfrac{1}{2}}$

$\therefore y(t) = -3e^{-\tfrac{1}{2}t}$

Exercise 20-5:

$y'' + 2y' + 2y = 2 \quad y(0) = 1 \text{ and } y(0)' = 0$

$\therefore s^2Y(s) - sy(0) - y(0)' + 2sY(s) - 2y(0) + 2Y(s) = \frac{2}{s}$

$\therefore (s^2 + 2s + 2)Y(s) = \frac{s^2+2s+2}{s}$

$\therefore Y(s) = \frac{1}{s}$

$\therefore y(t) = 1$

Exercise 20-6:

$\ddot{y} + \omega^2 y = 0 \quad y(0) = 0 \quad y(0)' = 1$

$\therefore s^2Y(s) - sy(0) - y(0)' + \omega^2 Y(s) = 0$

$\therefore (s^2 + \omega^2)Y(s) - 1 = 0$

$\therefore Y(s) = \frac{1}{s^2+\omega^2} = \frac{1}{\omega} \cdot \frac{\omega}{s^2+\omega^2}$

$\therefore y(t) = \frac{1}{\omega}\sin(\omega t)$

Exercise 20-7:

$f(t) = u(t-a) \text{ where } b = 1$

$\therefore F(s) = \frac{b}{s}e^{-sa} = \frac{1}{s}e^{-sa}$

$\therefore sF(s) - f(0) = e^{-sa} - 0 = e^{-sa}$

Exercise 20-8:

$Y(s) = -2a\frac{\left(a^2 - 3s^2\right)}{\left(a^2 + s^2\right)^3}$

Differential Equations in 24 Hours

Input Domain	Laplace Domain
sin(at)	$\dfrac{a}{a^2+s^2} = a(a^2+s^2)^{-1}$
-tsin(at)	$-a(a^2+s^2)^{-2}2s = 2as(a^2+s^2)^{-2}$
-t(-tsin(at))	$2a(a^2+s^2)^{-2} + 2as(-2)(a^2+s^2)^{-3}2s = \dfrac{2a(a^2+s^2) - 8as^2}{(a^2+s^2)^3}$ $= \dfrac{2a^3 - 6as^2}{(a^2+s^2)^3} = -2a\dfrac{a^2 - 3s^2}{(a^2+s^2)^3}$

Exercise 22-1:

$y''+4y'+13y=0$ with $y(0)=1$ and $y'(0)=2$

$S^2Y-sy(0)-y'(0)+4[sY-y(10)]+13Y=0$

$S^2Y-s+2+4[sY-1]+13Y=0$

$S^2Y-s+2+4sY-4+13Y=0$

$(Y)(s^2+4s+13)=s+2$

$Y=\dfrac{s+2}{s^2+4s+13}=\dfrac{s+2}{s^2+4s+4+9}$

$Y=\dfrac{s+2}{(s+2)^2+3^2}$

$y(t) = e^{-2t}\cos 3t$

Exercise 22-2:

$y'' + y = e^t$, $y(0) = 1$, $y'(0) = 1$

$s^2 Y - sy(0) - y'(0) + Y = \dfrac{1}{s-1}$

$s^2 Y - s - 1 + Y = \dfrac{1}{s-1}$

$(Y)(s^2 + 1) = s + 1 + \dfrac{1}{s-1}$

$= \dfrac{(s-1)(s+1) + 1}{s-1}$

$= \dfrac{s^2 - 1 + 1}{s-1}$

$(s^2 + 1) Y = \dfrac{s^2}{s-1}$

$Y = \dfrac{s^2}{(s^2+1)(s-1)}$

$\dfrac{s^2}{(s^2+1)(s-1)} = \dfrac{As}{s^2+1} + \dfrac{B}{s^2+1} + \dfrac{C}{s-1}$

$s^2 = As(s-1) + B(s-1) + C(s^2 + 1)$

$s^2 = As^2 - As + Bs - B + Cs^2 + C$

$s^2 = (A+C)s^2 + (B-A)s + C - B$

$A + \quad C = 1$

$-A + B \quad = 0$

$\quad -B + C = 0$

Differential Equations in 24 Hours

$$\begin{bmatrix} 1 & 0 & 1 & 1 \\ -1 & 1 & 0 & 0 \\ 0 & -1 & 1 & 0 \end{bmatrix}$$

$$\rightarrow \begin{bmatrix} 1 & 0 & 1 & 1 \\ 0 & 1 & 1 & 1 \\ 0 & -1 & 1 & 0 \end{bmatrix} \rightarrow \begin{bmatrix} 1 & 0 & 1 & 1 \\ 0 & 1 & 1 & 1 \\ 0 & 0 & 2 & 1 \end{bmatrix} \rightarrow \begin{bmatrix} 1 & 0 & 1 & 1 \\ 0 & 1 & 1 & 1 \\ 0 & 0 & 1 & 1/2 \end{bmatrix} \rightarrow \begin{bmatrix} 1 & 0 & 0 & 1/2 \\ 0 & 1 & 0 & 1/2 \\ 0 & 0 & 1 & 1/2 \end{bmatrix}$$

A=1/2

B=1/2

C=1/2

$$Y = \frac{s^2}{(s^2+1)(s-1)} = \frac{1}{2}(\frac{s}{s^2+1} + \frac{1}{s^2+1} + \frac{1}{s-1})$$

$$y(t) = \frac{1}{2}(\cos t + \sin t + e^t)$$

Exercise 22-3:

$y''+5y'+6y=0$ with $y(0)=0$ and $y'(0)=1$

$s^2Y-sy(0)-y'(0)+5sY-sy(0)+6Y=0$

$s^2Y-1+5sY+6Y=0$

$(s^2+5s+6)Y=1$

$$Y = \frac{1}{(s+2)(s+3)} = \frac{A}{s+2} + \frac{B}{s+3}$$

$1=A(s+3)+B(s+2)=(A+B)s+3A+2B$

$$\begin{bmatrix} 3 & 2 & 1 \\ 1 & 1 & 0 \end{bmatrix} \rightarrow \begin{bmatrix} 1 & 1 & 0 \\ 3 & 2 & 1 \end{bmatrix} \rightarrow \begin{bmatrix} 1 & 1 & 0 \\ 0 & -1 & 1 \end{bmatrix} \rightarrow \begin{bmatrix} 1 & 0 & 1 \\ 0 & 1 & -1 \end{bmatrix}$$

$$Y = \frac{1}{s+2} - \frac{1}{s+3}$$

$Y=e^{-2t}-e^{-3t}$

Appendix 1: Answers to Exercises

Exercise 22-4:

$y'' + 9y = S(t - \frac{3\pi}{2})$ $y(0)=0, y'(0)=0$

$s^2 Y + 9Y = e^{-\frac{3\pi}{2}s}$

$Y = \frac{1}{s^2+3^2} e^{-\frac{3\pi}{2}s}$

$Y = \frac{1}{3} \frac{1}{s^2+3^2} e^{-\frac{3\pi}{2}s}$

$Y = \frac{1}{3} \sin[3(t - \frac{3\pi}{2})] u(t - \frac{3\pi}{2})$

$= \frac{1}{3} \sin[3t - \frac{9\pi}{2}] u(t - \frac{3\pi}{2})$

$y = \frac{1}{3} \sin[3t - \frac{\pi}{2}] u(t - \frac{3\pi}{2})$

$\sin(A+B) = \sin A \cos B - \sin B \cos A$

$y = \frac{1}{3} [\sin 3t \cos \frac{\pi}{2} - \sin \frac{\pi}{2} \cos 3t] u(t - \frac{3\pi}{2})$

$= -\frac{1}{3} \cos 3t \, u(t - \frac{3\pi}{2})$

Exercise 22-5:

$L \frac{d^2 j}{dt^2} + R \frac{dj}{dt} + \frac{1}{c} j = -w e^{-wt}$

$j(0)=0, j'(0)=0$

$s^2 L J + s R J + \frac{1}{c} J = \frac{(-w)}{(s+w)}$

$J = \frac{1}{(Ls^2 + Rs + \frac{1}{c})} \frac{(-w)}{(s+w)}$

Differential Equations in 24 Hours

Exercise 22-6:

$$J = \frac{-1}{(s^2+2s+2)(s+1)}$$

$$\frac{-1}{(s^2+2s+2)(s+1)} = \frac{A}{(s+1)} + \frac{Bs+C}{(s^2+2s+2)}$$

$-1 = A(s^2+2s+2) + (Bs+c)(s+1)$

$-1 = As^2 + 2As + 2A + Bs^2 + Bs + Cs + C$

$-1 = (A+B)s^2 + (2A+B+C)s + 2A+C$

$$\begin{bmatrix} 1 & 1 & 0 & 0 \\ 2 & 1 & 1 & 0 \\ 2 & 0 & 1 & -1 \end{bmatrix} \rightarrow \begin{bmatrix} 1 & 1 & 0 & 0 \\ 0 & -1 & 1 & 0 \\ 0 & -1 & 0 & -1 \end{bmatrix} \rightarrow \begin{bmatrix} 1 & 0 & 0 & -1 \\ 0 & 1 & 0 & 1 \\ 0 & -1 & 1 & 0 \end{bmatrix} \rightarrow \begin{bmatrix} 1 & 0 & 0 & -1 \\ 0 & 1 & 0 & 1 \\ 0 & 0 & 1 & 1 \end{bmatrix}$$

$$\frac{-1}{s+1} + \frac{s+1}{(s+1)^2+1}$$

$y = -e^{-t} + e^{-t}\cos t$

$y = (\cos t - 1)e^{-t}$

Exercise 24-1:

$$\begin{cases} \frac{dy_1}{dt} = y_1 + y_2 \\ \frac{dy_2}{dt} = 4y_1 - 2y_2 \end{cases}$$

$\vec{y} = \begin{bmatrix} y_1 \\ y_2 \end{bmatrix}$, $A = \begin{bmatrix} 1 & 1 \\ 4 & -2 \end{bmatrix}$, $\vec{y} = e^{\lambda t}\vec{x}$, $A\vec{x} = \lambda\vec{x}$

Appendix 1: Answers to Exercises

$(A - \lambda I)\vec{x} = \vec{0}$, $\det(\lambda I - A) = \begin{vmatrix} (\lambda-1) & -1 \\ -4 & (\lambda+2) \end{vmatrix} = 0$

$(\lambda - 1)(\lambda + 2) - 4 = 0$
$\lambda^2 + \lambda - 6 = 0$
$(\lambda + 3)(\lambda - 2) = 0 \quad \Rightarrow \quad \lambda_1 = 2, \; \lambda_2 = -3$

For the $\lambda_1 = 2$ eigenvalue, solve for eigenvector $\vec{x}_1$:

$(\lambda_1 I - A)\vec{x}_1 = \vec{0} \quad \Rightarrow \quad \begin{bmatrix} 1 & -1 \\ -4 & 4 \end{bmatrix} \begin{bmatrix} x_{11} \\ x_{12} \end{bmatrix} = \begin{bmatrix} 0 \\ 0 \end{bmatrix} \quad \Rightarrow$

$\begin{bmatrix} 1 & -1 & 0 \\ -4 & 4 & 0 \end{bmatrix} \Rightarrow \begin{bmatrix} 1 & -1 & 0 \\ 0 & 0 & 0 \end{bmatrix}$

Let $x_{11} = s$. Then $x_{12} = s$. Therefore: $\vec{x}_1 = \begin{bmatrix} 1 \\ 1 \end{bmatrix}$

For the $\lambda_2 = -3$ eigenvalue, solve for eigenvector $\vec{x}_2$:

$(\lambda_2 I - A)\vec{x}_2 = \vec{0} \quad \Rightarrow \quad \begin{bmatrix} -4 & -1 \\ -4 & -1 \end{bmatrix} \begin{bmatrix} x_{21} \\ x_{22} \end{bmatrix} = \begin{bmatrix} 0 \\ 0 \end{bmatrix} \quad \Rightarrow$

$\begin{bmatrix} -4 & -1 & 0 \\ -4 & -1 & 0 \end{bmatrix} \Rightarrow \begin{bmatrix} 4 & 1 & 0 \\ 0 & 0 & 0 \end{bmatrix} \Rightarrow \begin{bmatrix} 1 & \frac{1}{4} & 0 \\ 0 & 0 & 0 \end{bmatrix}$

Let $x_{21} = s$. Then $x_{22} = -4s$. Thus: $\vec{x}_2 = \begin{bmatrix} 1 \\ -4 \end{bmatrix}$

$\vec{y} = c_1 e^{\lambda_1 t} \vec{x}_1 + c_2 e^{\lambda_2 t} \vec{x}_2 = c_1 e^{2t} \begin{bmatrix} 1 \\ 1 \end{bmatrix} + c_2 e^{-3t} \begin{bmatrix} 1 \\ -4 \end{bmatrix}$

Differential Equations in 24 Hours

$$\vec{y} = \begin{bmatrix} y_1 \\ y_2 \end{bmatrix} = \begin{bmatrix} c_1 e^{2t} \\ c_1 e^{2t} \end{bmatrix} + \begin{bmatrix} c_2 e^{-3t} \\ -4c_2 e^{-3t} \end{bmatrix}$$

$$\begin{bmatrix} y_1 \\ y_2 \end{bmatrix} = \begin{bmatrix} e^{2t} & e^{-3t} \\ e^{2t} & -4e^{-3t} \end{bmatrix} \begin{bmatrix} c_1 \\ c_2 \end{bmatrix}$$

Exercise 24-2:

$$\begin{cases} \dfrac{dy_1}{dt} = 4y_1 - 2y_2 \\ \dfrac{dy_2}{dt} = 5y_1 + 2y_2 \end{cases}$$

$$\vec{y} = \begin{bmatrix} y_1 \\ y_2 \end{bmatrix}, \quad A = \begin{bmatrix} 4 & -2 \\ 5 & 2 \end{bmatrix}, \quad \vec{y} = e^{\lambda t}\vec{x}, \quad A\vec{x} = \lambda\vec{x}$$

$$(A - \lambda I)\vec{x} = \vec{0}, \quad \det(\lambda I - A) = \begin{vmatrix} (\lambda - 4) & 2 \\ -5 & (\lambda - 2) \end{vmatrix} = 0$$

$$(\lambda - 4)(\lambda - 2) + 10 = 0 \implies \lambda^2 - 6\lambda + 18 = 0$$

$$\lambda = \dfrac{6 \pm \sqrt{36 - 72}}{2} = \dfrac{6 \pm \sqrt{-36}}{2} = 3 \pm 3i \implies \begin{array}{l} \lambda_1 = 3 - 3i \\ \lambda_2 = 3 + 3i \end{array}$$

For $\lambda_1 = 3 - 3i$ eigenvalue, solve for the eigenvector $\vec{x}_1$:

$$(\lambda_1 I - A)\vec{x}_1 = \vec{0} \implies \begin{bmatrix} -(1+3i) & 2 \\ -5 & (1-3i) \end{bmatrix} \begin{bmatrix} x_{11} \\ x_{12} \end{bmatrix} = \begin{bmatrix} 0 \\ 0 \end{bmatrix} \implies$$

Appendix 1: Answers to Exercises

$$\begin{bmatrix} (1+3i) & -2 & 0 \\ -5 & (1-3i) & 0 \end{bmatrix} \quad \text{multiply top row thru by } .1 - .3i$$

$$\begin{bmatrix} 1 & (-.2+.6i) & 0 \\ -5 & (1-3i) & 0 \end{bmatrix} \Rightarrow \begin{bmatrix} 1 & (-.2+.6i) & 0 \\ 0 & 0 & 0 \end{bmatrix}$$

$x_{11} = (.2 - .6i) x_{12}$ ⠀⠀ Let $x_{12} = s$, $x_{11} = (.2 - .6i)s$

$$\vec{x}_1 = \begin{bmatrix} .2 - .6i \\ 0 \end{bmatrix}$$

$$\vec{y} = e^{(3-3i)t} \begin{bmatrix} .2 - .6i \\ 1 \end{bmatrix} = e^{3t} e^{-3ti} \left(\begin{bmatrix} .2 \\ 1 \end{bmatrix} - i \begin{bmatrix} .6 \\ 0 \end{bmatrix} \right)$$

$$\vec{y} = e^{3t} (\cos(3t) - i \sin(3t)) \left(\begin{bmatrix} .2 \\ 1 \end{bmatrix} - i \begin{bmatrix} .6 \\ 0 \end{bmatrix} \right)$$

$$= e^{3t} \left(\begin{bmatrix} .2\cos(3t) \\ \cos(3t) \end{bmatrix} - i \begin{bmatrix} .6\cos(3t) \\ 0 \end{bmatrix} - i \begin{bmatrix} .2\sin(3t) \\ \sin(3t) \end{bmatrix} - \begin{bmatrix} .6\sin(3t) \\ 0 \end{bmatrix} \right)$$

$$= e^{3t} \left(\begin{bmatrix} .2\cos(3t) - .6\sin(3t) \\ \cos(3t) \end{bmatrix} - i \begin{bmatrix} .6\cos(3t) + .2\sin(3t) \\ \sin(3t) \end{bmatrix} \right) = \vec{q}_1 + i\vec{q}_2$$

$$\vec{y}_g = c_1 \vec{q}_1 + c_2 \vec{q}_2$$

$$= \frac{c_1}{5} \begin{bmatrix} e^{3t}(\cos(3t) - 3\sin(3t)) \\ 5e^{3t} \cos(3t) \end{bmatrix} - \frac{c_2}{5} \begin{bmatrix} e^{3t}(3\cos(3t) + \sin(3t)) \\ 5e^{3t} \sin(3t) \end{bmatrix}$$

Let $\dfrac{c_1}{5} \to c_1$ and $-\dfrac{c_2}{5} \to c_2$.

Differential Equations in 24 Hours

$$\begin{bmatrix} y_1 \\ y_2 \end{bmatrix} = \begin{bmatrix} e^{3t}(\cos 3t - 3\sin 3t) & e^{3t}(3\cos 3t + \sin 3t) \\ 5e^{3t}\cos 3t & 5e^{3t}\sin 3t \end{bmatrix} \begin{bmatrix} c_1 \\ c_2 \end{bmatrix}$$

Exercise 24-3:

$$\frac{dy_1}{dt} = -y_1 - y_2$$

$$\frac{dy_2}{dt} = 6y_1 + 4y_2$$

$y_1(0) = -2, \ y_2(0) = 6$

$$\vec{y} = \begin{bmatrix} y_1 \\ y_2 \end{bmatrix}, \ A = \begin{bmatrix} -1 & -1 \\ 6 & 4 \end{bmatrix}$$

$$\det(\lambda I - A) = \begin{vmatrix} (\lambda+1) & 1 \\ -6 & (\lambda-4) \end{vmatrix} = 0$$

$(\lambda+1)(\lambda-4) + 6 = 0 \Rightarrow \lambda^2 - 3\lambda + 2 = 0 \Rightarrow$
$(\lambda-2)(\lambda-1) = 0 \Rightarrow \lambda_1 = 1, \ \lambda_2 = 2$

For $\lambda_1 = 1$ find $\vec{x}_1$:

$$\begin{bmatrix} (\lambda_1+1) & 1 \\ -6 & (\lambda_1-4) \end{bmatrix} \begin{bmatrix} x_{11} \\ x_{12} \end{bmatrix} = \begin{bmatrix} 0 \\ 0 \end{bmatrix} \Rightarrow \begin{bmatrix} 2 & 1 \\ -6 & -3 \end{bmatrix} \begin{bmatrix} x_{11} \\ x_{12} \end{bmatrix} = \begin{bmatrix} 0 \\ 0 \end{bmatrix} \Rightarrow$$

$$\begin{bmatrix} 2 & 1 & 0 \\ -6 & -3 & 0 \end{bmatrix} \Rightarrow \begin{bmatrix} 2 & 1 & 0 \\ 0 & 0 & 0 \end{bmatrix} \Rightarrow \begin{matrix} 2x_{11} = -x_{12} \\ x_{11} = s \\ x_{12} = -2s \end{matrix} \Rightarrow \vec{x}_1 = \begin{bmatrix} 1 \\ -2 \end{bmatrix}$$

Appendix 1: Answers to Exercises

For $\lambda_2 = 2$ find $\vec{x}_2$

$$\begin{bmatrix} (\lambda_2+1) & 1 \\ -6 & (\lambda_2-4) \end{bmatrix} \begin{bmatrix} x_{11} \\ x_{12} \end{bmatrix} = \begin{bmatrix} 0 \\ 0 \end{bmatrix} \Rightarrow \begin{bmatrix} 3 & 1 \\ -6 & -2 \end{bmatrix} \begin{bmatrix} x_{11} \\ x_{12} \end{bmatrix} = \begin{bmatrix} 0 \\ 0 \end{bmatrix} \Rightarrow$$

$$\begin{bmatrix} 3 & 1 & 0 \\ -6 & -2 & 0 \end{bmatrix} \Rightarrow \begin{bmatrix} 3 & 1 & 0 \\ 0 & 0 & 0 \end{bmatrix}$$

$3x_{21} = -x_{22}$
$x_{21} = s \quad \Rightarrow \vec{x}_2 = \begin{bmatrix} 1 \\ -3 \end{bmatrix}$
$x_{22} = -3s$

$$\vec{y} = c_1 e^t \begin{bmatrix} 1 \\ -2 \end{bmatrix} + c_2 e^{2t} \begin{bmatrix} 1 \\ -3 \end{bmatrix}$$

$$\vec{y} = \begin{bmatrix} e^t & e^{2t} \\ -2e^t & -3e^{2t} \end{bmatrix} \begin{bmatrix} c_1 \\ c_2 \end{bmatrix}$$

Use initial conditions $y_1(0) = -2$, $y_2(0) = 6$

$$\begin{bmatrix} -2 \\ 6 \end{bmatrix} = \begin{bmatrix} 1 & 1 \\ -2 & 3 \end{bmatrix} \begin{bmatrix} c_1 \\ c_2 \end{bmatrix} \Rightarrow \begin{bmatrix} 1 & 1 & -2 \\ -2 & -3 & 6 \end{bmatrix} \Rightarrow \begin{bmatrix} 1 & 0 & 0 \\ 0 & 1 & -2 \end{bmatrix}$$

$c_1 = 0$
$c_2 = -2$ $\Rightarrow \vec{y} = -2e^{2t} \begin{bmatrix} 1 \\ -3 \end{bmatrix} \Rightarrow \vec{y} = \begin{bmatrix} -2e^{2t} \\ 6e^{2t} \end{bmatrix}$

Exercise 24-4:

$$\vec{y}' = \begin{bmatrix} -5 & -2 \\ 12 & 5 \end{bmatrix} \vec{y} \, , \, \vec{y}(1) = \begin{bmatrix} 1 \\ 0 \end{bmatrix}$$

Differential Equations in 24 Hours

$$\det(\lambda I - A) = \begin{vmatrix} (\lambda+5) & 2 \\ -12 & (\lambda-5) \end{vmatrix} = 0$$

$(\lambda+5)(\lambda-5) + 24 = 0 \Rightarrow \lambda^2 - 1 = 0 \Rightarrow \lambda_1 = 1, \lambda_2 = -1$

For $\lambda_1 = 1$ find $\vec{x}_1$:

$$\begin{bmatrix} (\lambda_1+5) & 2 \\ -12 & (\lambda_1-5) \end{bmatrix} \begin{bmatrix} x_{11} \\ x_{12} \end{bmatrix} = \begin{bmatrix} 0 \\ 0 \end{bmatrix} \Rightarrow \begin{bmatrix} 6 & 2 & 0 \\ -12 & -4 & 0 \end{bmatrix} \Rightarrow$$

$$\begin{bmatrix} 3 & 1 & 0 \\ 0 & 0 & 0 \end{bmatrix} \Rightarrow \begin{array}{c} 3x_{11} = -x_{12} \\ \text{let } x_{11} = -s \Rightarrow \vec{x}_1 = \begin{bmatrix} -1 \\ 3 \end{bmatrix} \\ x_{12} = 3s \end{array}$$

For $\lambda_2 = -1$ find $\vec{x}_2$:

$$\begin{bmatrix} (\lambda_2+5) & 2 \\ -12 & (\lambda_2-5) \end{bmatrix} \begin{bmatrix} x_{21} \\ x_{22} \end{bmatrix} = \begin{bmatrix} 0 \\ 0 \end{bmatrix} \Rightarrow \begin{bmatrix} 2 & 1 & 0 \\ 0 & 0 & 0 \end{bmatrix} \Rightarrow$$

$2x_{21} = -x_{22}$ let $x_{21} = -s, x_{22} = 2s \Rightarrow \vec{x}_2 = \begin{bmatrix} -1 \\ 2 \end{bmatrix}$

$$\vec{y} = c_1 e^t \begin{bmatrix} -1 \\ 3 \end{bmatrix} + c_2 e^{-t} \begin{bmatrix} -1 \\ 2 \end{bmatrix}$$

$$\begin{bmatrix} -e^t & -e^{-t} \\ 3e^t & 2e^{-t} \end{bmatrix} \begin{bmatrix} c_1 \\ c_2 \end{bmatrix} = \begin{bmatrix} y_1 \\ y_2 \end{bmatrix}$$

Use initial condition $\vec{y}(1) = \begin{bmatrix} 1 \\ 0 \end{bmatrix} \Rightarrow$

$$\begin{bmatrix} -e & -e^{-1} & 1 \\ 3e & 2e^{-1} & 0 \end{bmatrix} \Rightarrow \begin{bmatrix} -e & -e^{-1} & 1 \\ 0 & -e^{-1} & 3 \end{bmatrix} \Rightarrow$$

$$\begin{bmatrix} -e & 0 & -2 \\ 0 & e^{-1} & -3 \end{bmatrix} \Rightarrow \begin{bmatrix} 1 & 0 & 2e^{-1} \\ 0 & 1 & -3e \end{bmatrix} \Rightarrow$$

$$\bar{y} = 2e^{-1}e^{t} \begin{bmatrix} -1 \\ 3 \end{bmatrix} - 3ee^{-t} \begin{bmatrix} -1 \\ 2 \end{bmatrix}$$

$$\bar{y} = 2e^{t-1} \begin{bmatrix} -1 \\ 3 \end{bmatrix} - 3e^{1-t} \begin{bmatrix} -1 \\ 2 \end{bmatrix}$$

$$\bar{y} = \begin{bmatrix} -2e^{t-1} + 3e^{1-t} \\ 6e^{t-1} - 6e^{1-t} \end{bmatrix}$$

Exercise 24-5:

$$\bar{y}' = \begin{bmatrix} 1 & 2 \\ 2 & 1 \end{bmatrix} \bar{y}, \quad \bar{y}(-1) = \begin{bmatrix} 2 \\ 2 \end{bmatrix}$$

$$\det(\lambda I - A) = \begin{vmatrix} (\lambda-1) & -2 \\ -2 & (\lambda-1) \end{vmatrix} = 0$$

$$(\lambda - 1)^2 - 4 = 0 \Rightarrow \lambda^2 - 2\lambda - 3 = 0 \Rightarrow (\lambda - 3)(\lambda + 1) = 0$$

$$\lambda_1 = 3, \quad \lambda_2 = -1$$

For $\lambda_1 = 3$ find $\bar{x}_1$:

$$\begin{bmatrix} (\lambda_1-1) & -2 \\ -2 & (\lambda_1-1) \end{bmatrix} \begin{bmatrix} x_{11} \\ x_{12} \end{bmatrix} = \begin{bmatrix} 0 \\ 0 \end{bmatrix} \Rightarrow \begin{bmatrix} 2 & -2 & 0 \\ -2 & 2 & 0 \end{bmatrix} \Rightarrow$$

$$\begin{bmatrix} 1 & -1 & 0 \\ 0 & 0 & 0 \end{bmatrix} \Rightarrow \bar{x}_1 = \begin{bmatrix} 1 \\ 1 \end{bmatrix}$$

For $\lambda_2 = -1$ find $\bar{x}_2$:

$$\begin{bmatrix} (\lambda_2-1) & -2 \\ -2 & (\lambda_2-1) \end{bmatrix} \begin{bmatrix} x_{11} \\ x_{12} \end{bmatrix} = \begin{bmatrix} 0 \\ 0 \end{bmatrix} \Rightarrow \begin{bmatrix} -2 & -2 & 0 \\ -2 & -2 & 0 \end{bmatrix} \Rightarrow$$

$$\begin{bmatrix} 1 & 1 & 0 \\ 0 & 0 & 0 \end{bmatrix} \Rightarrow \bar{x}_2 = \begin{bmatrix} 1 \\ -1 \end{bmatrix}$$

$$\bar{y} = c_1 e^{3t} \begin{bmatrix} 1 \\ 1 \end{bmatrix} + c_2 e^{-t} \begin{bmatrix} 1 \\ -1 \end{bmatrix}$$

$$\bar{y} = \begin{bmatrix} e^{3t} & e^{-t} \\ e^{3t} & -e^{-t} \end{bmatrix} \begin{bmatrix} c_1 \\ c_2 \end{bmatrix}$$

Use initial condition:

$$\begin{bmatrix} e^{-3} & e \\ e^{-3} & -e \end{bmatrix} \begin{bmatrix} c_1 \\ c_2 \end{bmatrix} = \begin{bmatrix} 2 \\ 2 \end{bmatrix} \Rightarrow \begin{bmatrix} e^{-3} & e & 2 \\ e^{-3} & -e & 2 \end{bmatrix} \Rightarrow \begin{bmatrix} e^{-3} & e & 2 \\ 0 & -2e & 0 \end{bmatrix} \Rightarrow$$

$$\begin{bmatrix} e^{-3} & 0 & 2 \\ 0 & 1 & 0 \end{bmatrix} \Rightarrow \begin{matrix} c_1 = 2e^3 \\ c_2 = 0 \end{matrix} \Rightarrow \bar{y} = 2e^3 e^{3t} \begin{bmatrix} 1 \\ 1 \end{bmatrix} = 2e^{3t+3} \begin{bmatrix} 1 \\ 1 \end{bmatrix}$$

$$\bar{y} = \begin{bmatrix} 2e^{3t+3} \\ 2e^{3t+3} \end{bmatrix}$$

Appendix 2: Timeline

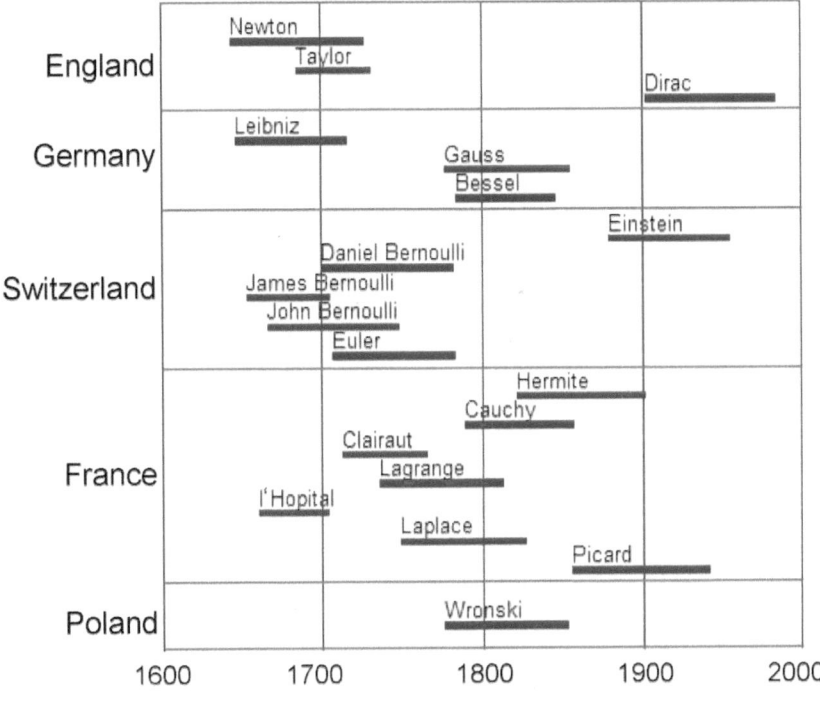

Index

A

advancing subscripts	Hour 18
arbitrary constant	Hour 1
auxiliary equation	Hours 12, 13, 14, 17

B

basis for the solution space	Hour 12
basis functions	Hours 12, 16
basis vectors	Hours 12, 15
beer froth rate of disappearance	Hour 1
bell curve	Hour 5
Bernoulli brothers	Hour 7
Bernoulli, Daniel	Hour 18
Bernoulli, John	Hour 3
Bernoulli's equation	Hour 10
Bessel functions	Hour 18
Bessel's equation	Hour 18
boundary conditions	Hours 5, 12
boundary value	Hour 5

boundary value problem (BVP)	Hours 5, 12

C

candidate solution	Hours 4, 15
Cauchy, Augustin	Hour 17
Cauchy-Euler equation	Hour 17
chain rule	Hours 3, 7, 14
circuit	
discrete-time	Hour 4
LC	Hour 12
LRC	Hours 21, 22
RL	Hour 8
Clairaut, Alexis Claude	Hour 7
Clairaut's condition	Hour 7
Clairaut's Theorem	Hour 7
Classifying equations and solutions	Hour 4
closed under differentiation	Hour 15
coefficient generator	Hour 18
compatible powers	Hour 18
complex conjugate roots	Hours 13, 17
complex function	Hour 11
complex numbers	Hour 11

Index

conjugate	Hour 11
conjugate of a complex function	Hour 11
constant coefficient homogeneous system	Hour 23
constant function	Hour 7
continuity	Hour 6

D

degrees of freedom	Hour 4
derivative of step function	Hour 20
derivatives, Laplace transforms of	Hour 19
difference equation	Hour 4
differential equation	Hour 1
linear	Hour 4
nonlinear	Hour 3
differential	Hour 1
exact	Hour 7
differentiation, closed under	Hour 15
differentiation via multiplication	Hour 19
digital signal processing	Hour 4
dilation equations	Hour 4
Dirac delta function	Hour 21
Dirac, Paul A. M.	Hour 21

discontinuous functions	Hour 19
discontinuous square wave	Hour 19

E

eigenvalue	Hour 23
eigenvalue equation	Hour 23
eigenvector	Hour 23
Einstein, Albert	Hours 1, 5
Einstein's postulate	Hour 5
Einstein's Special Theory of Relativity	Hours 2, 3, 5, 6
Einstein's Theory of Heat Capacity	Hour 1
elastic string fixed at both ends	Hour 12
electrical resonance	Hour 12
electromotive force	Hour 8
elementary row operation	Hour 23
Euclidean two-space	Hour 12
Euler, Leonhard	Hour 11
Euler's formula	Hour 11
exact differential	Hour 7
exact equation	Hour 7
exactness	Hour 7
existence of solutions	Hour 6

Index

explicit solution	Hour 2
exponential decay equation	Hours 2, 3

F

family of solutions	Hours 2, 4, 15
first-order linear systems with constant coefficients	Hour 23
first translation theorem	Hour 21
Fourier Transform	Hour 11
functional coefficients	Hour 16

G

Gaussian integers	Hour 11
general form	Hour 4
general solution	Hours 5, 12
guitar string	Hour 12

H

half-life	Hour 2
harmonic oscillator	Hours 2, 3, 4, 5, 12, 13, 19, 21
harmonic oscillator, struck	Hour 21
harmonics	Hour 12
heat equation	Hour 4

helper function	Hours 7, 9
Hermite, Charles	Hour 18
Hermite functions	Hour 18
Hermite's equation	Hour 18
higher-order differential equations	Hour 12
higher-order linear differential equations	Hour 12
homogeneity	Hour 12
homogeneous coefficients (method of)	Hour 9
homogeneous linear differential equation	Hour 12
homogeneous linear ODE's with constant coefficients	Hour 12
homogeneous of degree α	Hour 9
homogeneous solution	Hour 12
Homogeneous solution table	Hour 17

I

identity matrix	Hour 23
imaginary numbers	Hour 11
implicit solution	Hour 4
impulse of unit area	Hour 21
impulse response	Hour 21
indefinite integral	Hour 1
independent basis vectors	Hours 12, 15

Index

initial conditions	Hour 12
initial value	Hour 4
initial value problem (IVP)	Hour 4
inner product	Hour 11
inner product between functions	Hour 11
input domain	Hour 19
integrating factor	Hour 8
inverse-Laplace transform	Hour 19
inverse transform	Hour 19

K

Kirchoff's Law	Hours 8, 12, 21, 22

L

LC circuit	Hour 12
LRC circuit	Hours 21, 22
Lagrange, Joseph-Louis	Hour 16
Laplace domain	Hours 19, 21
Laplace, Pierre-Simon	Hour 19
Laplace Transform	Hours 4, 19, 20, 21, 22
Laplace Transform translation theorems	Hours 21, 22

Laplace transforms of derivatives	Hour 19
Leibniz differential notation	Hours 3
Leibniz differentials	Hours 1, 3
Leibniz, Gottfried	Hours 2, 19, 20, 21, 22
Leibniz's rule	Hour 19
linear algebra	Hour 23
linear dependence	Hour 12
linear differential equation	Hours 4, 8, 12, 23
$\quad n^{th}$ –order	Hour 12
linear differential equations of the first order (method of)	Hour 8
linear equation	Hours 4, 8
linear independence	Hour 12
linear operator	Hour 18
linearity of the Laplace transform	Hour 19
linearly independent basis vectors	Hour 12
Lorentz contraction	Hour 2
Lorentz Transformation	Hour 2

M

Maclaurin expansion	Hour 18

N

n-parameter family of solutions	Hour 4
neutron decay	Hour 2
Newton, Isaac	Hour 1
Newton's fly speck notation	Hour 4
non-constant coefficients	Hours 16, 18
non-homogeneous equation	Hour 12
non-homogeneous linear ODE with constant coefficients	Hour 15
non-trivial solution	Hours 14, 23
normal form	Hour 4
norm of a complex number	Hour 12
norm of a function	Hour 12
notation	
dot	Hours 2, 4
fly speck	Hour 4
primed	Hour 1

O

order of a differential equation	Hour 4
ordinary differential equations	Hours 4, 12
orthogonal functions	Hour 11

orthogonal vectors	Hour 12

P

partial differential equations	Hour 4
particular solution	Hours 4, 6, 12
Picard, Émile	Hour 6
Picard's Theorem	Hour 6
piecewise defined step	Hour 19
Poisson	Hour 19

R

radio	Hour 12
RL circuit	Hour 8
rational functions	Hour 19
ratios of polynomials	Hour 19
recursion formula	Hours 4, 18
reduction of order	Hours 13, 14
relativity equation	Hours 3, 5
relativity, special	Hours 2, 5
reverse transform	Hour 19

S

second translation theorem	Hour 21

Index

separable variables	Hour 3
series solution	Hour 18
solution	Hours 1, 2
existence of	Hour 6
explicit	Hour 2
general	Hours 5, 12
homogeneous	Hour 12
implicit	Hour 4
non-trivial	Hours 14, 23
particular	Hours 4, 6, 12
trivial	Hour 4
uniqueness of	Hour 6
solution space	Hour 12
basis for the	Hour 12
solving a differential equation	Hours 1, 2
space capsule	Hours 2, 5
Special Relativity	Hours 2, 5
spring	Hours 1, 4, 5
standard form	Hour 16
step function, derivative of	Hour 20
string	
vibrating	Hours 3, 5, 18

elastic	Hour 12
struck harmonic oscillator	Hour 21
subscripts, advancing	Hour 18
superposition principle	Hour 12
system of n first order linear differential equations	Hour 23
systems	Hours 1, 23, 24

T

transform domain	Hour 19
transform pair	Hour 19
translation theorem, first	Hour 21
translation theorem, inverse first	Hour 21
translation theorem, second	Hour 21
translation theorems	Hour 21
trivial solution	Hour 4
true sentence	Hour 2

U

undetermined coefficients	Hour 15
uniqueness of solutions	Hour 6
unit step function	Hour 19

Index

V

variation of parameters	Hour 16
vector space	Hour 12
verifying solutions of differential equations	Hour 2
vibrations of displaced atoms	Hour 1
vibrating string	Hours 3, 5, 18

W

wavelet	Hour 4
Wronski, Hoëné	Hour 16
Wronskian	Hour 12

Scott Imhoff, PhD

Dr. Scott Imhoff received an MS in physics from Purdue University and a PhD in mathematics from University of Maryland Baltimore County. He is a Senior Principal Engineer at Raytheon Company in Aurora Colorado and a Lecturer in the Department of Mathematical and Statistical Sciences at University of Colorado Denver.

www.ingramcontent.com/pod-product-compliance
Lightning Source LLC
Chambersburg PA
CBHW020733180526
45163CB00001B/221